NATURAL PEST CONTROL

Alternatives to Chemicals
for the
Home and Garden

by
Andrew Lopez
The Invisible Gardener

Published by
Harmonious Technologies

ISBN 0-9629768-4-9

By Andrew Lopez
The Invisible Gardener
P.O, Box 4311
Malibu, CA 90264
Telephone: (310) 457-4438
Fax: (310) 457-5003
E-mail: andy@invisiblegardener. com
Web page site: http://www.invisiblegardener.com
Artwork/layout Cindy Barry

Published by Harmonious Technologies
P.O. Box 1716
Sebastopol, CA 95473
Telephone: (707) 823-1999
Fax: (707) 823-2424
E-mail: info@homecompost.com
Internet: www.homecompost.com

The price of *Natural Pest Control: Alternatives to Chemicals for the Home and Garden* is $19.95 plus shipping and handling fees of $3.50 US, $5.50 all other countries. An order form has been provided on the last page.

FOREWARD

It is my pleasure to be able to introduce to you this manual of organic information for the home and garden, farmer and professional, written by an unusually well-informed and qualified person, Andrew Lopez, or as he is better known, "The Invisible Gardener". We all know by this time that our planet is in trouble, and that we are the ones who must do what is necessary to save her. Each one of us can work on our small piece of the world to set up a viable eco-balance system. Understanding the alternatives to chemicals is an important step in this process. It is necessary for us to realize that a change in the way we treat our own environment can make a big difference in the general eco picture. Chemical fertilizers and pest controls create enormous damage to the environment (pollution does begin at home) because they sterilize wherever applied (meaning they kill off all of the favorable bacteria necessary for a working eco-system), contaminate ground waters and kill fauna. Everything that we have learned to do with chemicals can be done with the organic system. The Invisible Gardener gives you all the information you need to do it, starting with the recipes (formulas), adding his unique methods, and giving you the sources of whatever you need.

The Organic system is not something to be used only in the vegetable garden, or to control ants, or to avoid toxic products. As expressed by The Invisible Gardener in this splendid book, it means total commitment to healing the good earth naturally. This often means proper fertilization and foliar feeding so that plants can protect themselves. It also means new concepts for tree management, flower propagation and even compost making, plus special remedies for problems inside the home and out.

The methodology is all there how to switch away from toxic technology, and how to avoid pollution of air, land and water. I especially like being able to use, say, Tabasco sauce to control pests eating my vegetables, or baking soda for my roses, or knowing how to use rock dust on my trees, roses, gardens, and lawns.

Andy's knowledge extends into the house, too. One of my favorite chapters is "Dances with Ants." He also saved me a great deal of money in helping me with carpenter ants (I thought they were termites!) and with brown recluse spiders, which drop in from time to time.

It's all so easy, safe and economical, too!

We know how to make our planet into a Garden of Eden. It is time we do it. Here is a tool. I recommend we all use it. Each one of us should have our own Green Acres.

Yours Organically,

Eddie Albert

Chief Seattle was not talking about the World Wide Web when he made the statement below but meant instead "linked together", like the spider's web. Everything we do to the earth effects us all. We cannot escape it's effects any more than we can escape from the chemicals that we are using on this planet!

"Man did not weave the web of life. He is merely a strand in it.
Whatever he does to the web, he does to himself."
Chief Seattle 1856

People want to be able to say, "I am helping the earth by not using any chemicals that pollute her." Actions and words are not the same. People do far more damage to the earth than they realize!

For many years I have been saying, "Be careful of what you use and what goes into the earth". Even if chemical companies say it's safe and that it breaks down into harmless compounds, do not believe them! Scientists are finding that these harmless compounds do indeed get together with other so called harmless compounds to form much deadlier compounds, which not only harm humans but all wildlife they meet!

Researchers say substances that have little effect on sex hormones individually become much more potent when combined! The unexpected results mean safety standards are inadequate. Does this sound familiar? Animals in the wild are being effected at a tremendous rate. Animals like the Florida alligator and Great lakes' birds are being born with exceedingly high estrogen levels. These animals are being born sometimes with both male and female organs! They cannot mate and their line (species) will die!

This is due to chemicals that we use; from pesticides, to fertilizers, to everyday household items, to the medicine we use. All chemicals can have disastrous effects on wildlife and on humans as well! We must learn to control the use of these chemicals. We must learn to use organically safe chemicals that Mother Nature herself has tested for us. We must stop using chemicals that pollute ourselves, our children, our wildlife before it's too late. So the next time you reach for that spray bottle of ———, think about it. Do you have to use this? Is there another way?

This book is dedicated to the goal of helping you to choose the organic method over the chemical method.

It can be done 100% organically. It can be done and it does work! Please do not give in to the chemicals (did you know that beer has over 72 chemicals in it)?! Think of your children and the mess we are leaving them to handle. Think of the wildlife we are destroying by using these chemicals. Think about the ingredients of what you want to buy before you buy it! Ask yourself do I know what this stuff will do to the earth? If you do not have an answer, don't use it! If you do not buy it, they won't make it!

By following the 5 R's...Reduce, Reuse, Recycle, Replant, and Rethink; we will help the Earth to Heal herself and us!

Happy Growing Organically of course!

The Invisible Gardener of Malibu
Since 1972

Visit our web site at: http://www.InvisibleGardener.com

iv

TABLE OF CONTENTS
Foreward by Eddie Albert

1

Dances With Ants

How to Control Ants Organically

About the Ant

Ants are one of the most familiar insects in the world. It has been said that there are more ants on earth than there are stars in the visible sky—but who counts that high? I have also heard that it takes approximately 500,000 ants to add up to one pound, yet the combined weight of all the ants on earth is greater than the combined weight of all humans on earth. Also, ants outnumber all other terrestrial animals and live almost anywhere on earth they want. In other words, there are a lot of ants out there!

Ants can lift from 6 to 10 times their body weight for a distance of 300 times their own length even uphill and over obstacles, making them the strongest creatures on the planet—in relation to size. Combine their super strength with their advanced intelligence and advanced social structure, and you may understand why the ants have not changed much in their basic structure in the last few million years, even though other creatures have. It is their near-perfect form that allows ants to live happily in many different types of habitats, from the forest to the city. They develop into different species according to their environment.

A few of the more interesting ants are the thief ants which live in the walls of larger ants' nests as mice live in the walls of our houses; the umbrella ants, which cut leaves and carry them over their heads like umbrel-las as they travel around; and the leaf cutters, which cut leaves, roll them into balls and allow fungus to grow on them. Ants are just as comfortable in an apartment as in a swamp, provided there's plenty of food and water. Ants have even been found aboard airplanes, on boats in the middle of the ocean, and in speeding trains. While ants do not eat much plant material they are among the main predators of living creatures, from other insects to animals, birds, etc. In Africa, ants have been known to bring down an elephant for food, by climbing into its ears and eating it from the inside out. Some ants spit out chewed leaves in their underground farms to feed their underground gardens of fungi. They will raise fungi throughout the year, providing their colony with an endless food source.

Watching a single ant gives one only a slight idea of the group intelligence behind it, directing work that is being carried out for the whole. One ant may be busy dragging a grasshopper much bigger than itself, while another ant is following a scent trail that will lead it to a food or water source. That ant goes from place to place, marking the trail for others to read and follow, while another is in the midst of a battle. They know exactly where your kitchen is! Ants are constantly cleaning themselves with their antennae. They have developed a very sophisticated system of communication. Whenever their colony needs anything, the call goes out and the workers tell the scouts what is needed and off they go looking for just that food. When they find it they relay the message back to the colony and soon every one in the colony will know it, and off go the workers, in force, to get it and bring it back.

Ants are survivors. They recover very quickly from any disaster, especially those of man's making. They have actually adapted to benefit from mankind. They have even learned how to train us. They know how we will react to any given situation and know what buttons to push. They have adopted us as they have adopted the aphids! They remember events and places for several months. Just leave a piece of food or some sweets to be found by them and they will have your place mapped and marked with special paths that lead right to it. It is for this reason that you will not be able to eliminate ants entirely from your environment, and that you would be wise not to do battle with them, but to communicate with them and to reach a mutual understanding.

Ants belong to the hymenoptera family of insects (like bees and wasps), and to the formicidae family, which means simply that they live in colonies. In each colony there are three classes of individuals: the females or queens (fertile females), males (used for mating only), and the workers (wingless sterile females). The workers are the busy members of the family and are highly endowed (hmm?) with brain power and mechanical skills. The workers are further divided into three classes: the main workers, minor workers (gardeners), and the soldiers. The main and minor workers differ in size. The soldiers have huge heads and powerful jaws. They can be seen guarding the nest and defending it against attacks from other ants, etc. Worker ants can live from one to four years while some queens live up to 20 years. The homes of the ants differ greatly, depending upon the kind of ant and the terrain they have to deal with.

Stages of the Ant Colony

There are several stages the ants go through in establishing their colony, which is important to know in order to discourage them from moving into your household.

IN THE BEGINNING

Before they swarm, the males and queens are the aristocratic members of the family and are treated as such. They are kept clean and fed well. Their heaviest labor is a mock battle or a game of tag: You're it!

IT'S PARTY TIME!

Swarming, the first stage, starts with the nuptial flight, when the virgin females leave the colony of their parents and take with them some of the males; both are winged. They explore the surrounding neighborhood and perhaps some distant suburbs looking for a nice place to start a new home. There are far more females than males. Many die within the first hour, falling prey to birds, spiders and other predators. The females fly into swarms of males and mate, often with more than one male, in mid air! After mating, the inseminated females continue the search for locations for their new colony, while the males (no longer needed) either die at the hands of their lover or, if lucky, manage to leave the queen and survive for up to three days.

The Queen after fertilization

Male ant dies shortly after mating

Sterile female worker tends larvae

The queen has to either rejoin her old colony, join another colony or start a new colony. Alternatively, the queen can attack another colony and kill its current queen, taking over her colony. Once she has found her new location, she will drop off her wings, dig herself a nest, and seal herself in to lay her eggs. Depending on the species of ant, the queen will either seal herself off permanently in total isolation (called claustral), or she will forage outside for food (called partially claustral). There are two types of swarming: budding, when workers leave the main colony with one or more queens and start a new nest, and fission, where parts of the colony containing fertile queens separate, each queen taking with her a large portion of workers.

THE ERGONOMIC STAGE

The next stage of colony growth is called the ergonomic stage, in which the first workers born develop into mature workers. These first workers are usually small due to shortages of food and will get larger with each generation as more food becomes available to the colony. The main purpose of this stage is for colony growth. Here the queen will lay only fertilized eggs, which become females. Whether they become workers, soldiers, or new queens depends on the care they will get and upon the chemical signals they receive from their nurses.

THE REPRODUCTIVE STAGE

Next comes the reproductive stage, when the colony begins to produce males and fertile females. This is where the queen produces unfertilized eggs that mature into males and, at the same time, the workers will begin to groom female eggs into new queens.

When these queens have matured, it is time for the cycle to begin again with the nuptial flight.

Ants and Aphids

Ants have very advanced societies, complex and highly organized. Some ants live on household foods, sweets or protein, some eat only aphid honeydew, some at seeds, grains or vegetable roots, some grow their own food such as fungi, and of course most ants eat other insects.

Ants help to spread various types of bacterial diseases through their "farming" practices of herding aphids from one part of the plant to another. Ants will occasionally charge into a group of aphids. To the ants, these aphids are a great prize. The ants will stroke the aphids with their antennae until the aphids secrete a drop of a sweet liquid known as honeydew; this the ants store in their second stomachs. They do this until the second stomach is full; only then will an ant ingest the honey dew for itself, into its first stomach. The ant takes the honeydew back to the nest where it is delivered first to the nursery to feed the ant babies, and then to any other needy ants in the nest. The nectar is, unfortunately, not cleaned up very well after a meal, and the leftover nectar becomes food for many different types of bacteria and viruses. It also attracts many other insects such as whiteflies, scales, mites, and so on.

For ants, aphids are valuable food producers. They cannot be left

"Ants are involved in more than 70 percent of the pest problems associated with both the home and garden."

alone because other colonies might also find them, so they start to keep the aphids nearby, inside a corral built especially for them deep in the nest's tunnels. The aphids are happy slaves and will not only produce food for the ants, but help in digging out the tunnels as well. The ants will even care for the aphid babies as they do their own. Not all aphids have this relationship with the ants. Most aphids are wild and belong to none. The ants have to capture them and train them, much like we do with horses. When they need more aphids for the herd they go out and hunt the wild aphids.

Ants have relationships with many different creatures in the garden. They are involved in everything that happens in the garden.

The No-Pollution Solution

If you employed one of the ubiquitous services that claims to solve ant problems with chemicals, every few months someone would come by and spray your property, put up the now famous "Do not walk on the grass for 48 hours" sign, and that would be that.

There are many problems that arise from employing this method of pest control. It pollutes the land, groundwater, and produce. Numerous health hazards arise that result from exposure to these chemicals. And, most importantly, this method only increases the ants' resistance. Ants quickly develop an immunity to all chemicals, thus requiring stronger chemicals the next time around.

The solution to these problems is straightforward: Stop using chemicals and use a different approach. Viable alternatives exist to chemical fertilizers, pesticides, and herbicides that kill organisms within the soil, thereby upsetting its natural balance.

There are two ways to sterilize the soil: by using heat and by using chemicals.

When used in the soil, chemicals completely sterilize it, killing off the beneficial organisms (called simply "beneficials") that live in it. With the beneficials unable to reestablish themselves, predators gain the upper hand. Predators will always establish themselves first! Chemicals destroy the balance; when this natural balance is upset, the natural systems break down, causing infestations of pests or disease to occur.

NATURE'S GARDENERS

Ants play a very important role in the plant kingdom. They are not pests, but in fact nature's gardeners. When kept within their boundaries, they perform many functions which are important not only to the plant kingdom, but by extension, to all living things. Ants reveal to us the importance of nature's balance. They serve as indicators that tell us when there is balance and when there is not. It is unnatural and impossible to eliminate ants from the face of the earth. Remove them and a great deal of damage will occur. The answer lies in learning to control them. To keep ants off your roses, out of the house, and off the fruit trees, you will have to understand the various relationships that ants have with various plants, animals, and humankind. Timing is important. Ants help pollinate fruit trees and flowers. Thus, they must be allowed on the plants during this time for proper pollination, but not at other times, such as when the fruit trees are bearing. It is during this time that you use barriers to keep them off.

Remember—One of the best ways to deal with the ant problem is to realize that properly fed and healthy plants are less attractive to ants, other insects, and diseases than plants that are sick, weak and under stress due to improper feeding and watering.

Providing your plants 100 percent organic natural fertilizers is a very important step in reducing plant stress. Using compost and natural fertilizers helps keep the soil alive. It is this life that provides plants with the nutrition they need to keep insects at bay. The greater the nutritional levels, the lower the stress levels. Another way to improve plants' health is to use liquid seaweed to spray the leaves. It is best to blend different types of liquid seaweed to obtain a complete collection of. all the trace minerals needed for healthy plant growth. Water your plants regularly. Set up a watering schedule where conservation is required and stick to the schedule. If you are able to, use drip systems.

MULCHING

Mulching is very important as ants love dry, infertile soil. The importance of compost cannot be overstated. Compost properly made will provide an important source of nutrients and bacteria, needed by the plants for proper growth. (See the compost chapter for more info.) Mulching helps the soil retain moisture longer allowing for more biodiversity in the soil to occur.

Ants know when plants are stressed, and act accordingly. They have been around much longer than we have. The land your house is on was inhabited by ants long before it was inhabited by you. The earth is their home as well as ours. Don't try to fool them, you'll only fooling yourself. Don't be in a hurry—these things will take time.

Follow The Yellow Brick Road

Ants build what could be described as underground condominiums, and use "highways" to get from one place to another. They don't have to build these superhighways, since they prefer using paths that humans, other animals, and other insects have already established. They choose the path of least resistance, following sprinkler lines, drip lines, sidewalks, hose lines, or any other route that gets them where they are going with the least amount of effort.

When using natural sprays and barriers you must first find these paths, then spray or place the barriers so the ants will be routed away from where you do not want them to go. You will find that it helps to provide food and water in a different location. This should be done in an area where you don't mind their presence. Remember, you will not be able to completely rid your property of all ants. Even if you did, new ants would simply move in before you know it. The key to dealing with ants successfully is a balanced environment. Nature plays by certain rules. Respect these rules and they will respect you. Instead of imposing your own rules on nature, learn how to achieve a balanced environment within your own ecosystem. Nature will naturally keep ants in line, provided you help her out. Since we are part of nature, you might say she's helping herself. Give her time to regain her balance. One year should

The Invisible

Gardener Says:

_____

"If you can control the ants, you will have a greater chance of controlling your general insect problems and plant diseases, provided you maintain a healthy, balanced environment. Spraying a chemical pesticide, even one that rids you of your ant population for a few weeks or even months, does not solve your ant problem."

suffice, but the important thing to remember is that it will take time to control ants. It will be time well spent and the results will last forever.

Manufactured Demand

We do more damage to our environment than ants ever could. The chemicals we use to kill ants will never work with any real success because ants will always return eventually, stronger then ever. Many of the chemicals used to kill ants are leftovers from the stockpile of chemical warfare weapons developed during WW II. Newer, more deadly strains of these substances are developed every year to meet the demands of chemical farmers. Chemicals applied directly to the soil will purge it of all living things. Once sterilized in this manner, soil will not support life. Dead soil is a non-functioning system. Plants growing in it will experience great stress and eventually die.

The Six Steps to Controlling Ants

STEP 1....RECON AND SHOPPING LIST

Recon

The very first step in controlling ants around your home is for you to take a walk around and see if you're really cleaning up after yourself as well as you can. Ants are very smart - they know a good thing when they see it. If you are providing food and water for them then they certainly will know it and come around looking for it.

Don't leave out food overnight; always dispose of garbage properly. Usually the areas where the ants are seen in the house are near sources of either food or water or both. Do not provide a water source for them. Leaky water systems, such as hose connections, are a ready source of water and an ant colony will often be set up near them.

Are your children helping? What do you do with your pet food? Another important aspect of Step 1 is to locate any ants that are coming into your house and find out how they are coming in. Follow them around and see if they are entering through holes or cracks in the wall.

Let's go over what you should have done by now:

☑ cleaned up

☑ controlled sources of water

☑ observed how they enter and exit house

☑ observed what type(s) of food they are going after

☑ observed any trails around the outside of the house and where they are coming from

☑ walked around your property and tried to locate current ant colonies

Shopping List

☑ Dr. Bronners Peppermint Soap . You can get this soap at any health food store or food co-op. Dr. Bronners also makes Almond Oil Soap and several others. I would pick up one of each; rotating them pro duces the best results. You can also use Jungle Rain, if you prefer.

"Ants build what could be described as underground condominiums, and use "highways" to get from one place to another."

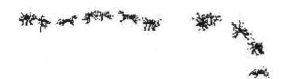

☑ 4 finch bird nesting houses - your local pet store - and 16 wood screws

☑ 8 oz honey or white sugar or corn syrup

☑ 8 oz plastic cups—one for each bird house

☑ a quart sprayer (or gallon sprayer)

☑ a caulking gun and caulk (or chewing gum)

☑ boric acid

☑ instant grits or yeast

Step 2.....On the Ant Trail

Dr. Bronners Peppermint soap: In your quart sprayer (gallon sprayer OK), mix the Dr. Bronners Soap at the rate of 1 tablespoon per quart water. Starting in center of house, spray the areas ants are seen in. Spray in the bathrooms and in the kitchen as well as on the floors and walls, especially in any holes or cracks where they were seen entering and exiting. (It's also okay to add the soap to mop water.) Spray also around the outside of the house where the wall meets the ground and at any other points where the ants are coming in. They follow electrical wiring, water pipes, air conditioning ducts, etc. See also Organic Ant Sprays, later in chapter.

Caulking gun and caulk: Once you have sprayed the soap then you caulk up all entrances and exits that the ants were using. Any type of material can be used to close these entrances. I sometimes use chewing gum! Elmers glue is OK to use if you don't mind the smell. Any type of natural compound can be used.

4 finch bird nesting houses: Once you have sprayed and then closed off all known entrances to the house, then you are ready to construct your Ant Cafes.

4-8 oz plastic cups: Inside each Ant Cafe you will place one 8 oz plastic cup. This is where you will place food and or water for the ants to eat. (You can use any other type of container that fits inside the ant cafes; I have used cat or dog food containers, glass bottles, etc.)

8 oz honey or white sugar or corn syrup: In the 8 oz cup mix 4 oz water with 1 oz honey or corn syrup or ¼ cup white sugar. Stir. We start with the sugar here, but it is best to start with whatever the ants were going after when you observed them in step 1. The Ant Cafe repast must be changed every month to insure the that ants don't lose interest. Do not add any boric acid until 2 weeks have gone by and the ants are actively going after their new food supply.

Boric acid: Boric acid is what you will add to the food to kill the ants. Please read the section "Killing or Not Killing Ants". Boric acid is sold at most garden centers. Ortho sells a clean boric acid as well ant control brands such as Drax and Terro

Alternative Sprays: There are some products that you can use to spray with that are natural and safe when used as directed. Inside House: Citra Solve is a natural citrus based cleaner and not actually a soap. It has a pleasant citrus smell but if you are sensitive to odors then you should take a light sniff before you buy it. This stuff will dissolve most insect bodies

Finch Bird Nesting House

Once you have sprayed and then closed off all known entrances to the house, then you are ready to construct your Ant Cafes.

"A big step in controlling ants is understanding how ants use trails to find their way around. Once you have started to keep the ants out of your house you must continue to fool them into staying out of your house. The Ant Cafe is an important tool towards this end."

it comes into contact with yet it is safe to use in the home and garden. Use ½ tablespoon per quart water. Any Herbal liquid soaps such as Green Ban Dog Shampoo, with its many different Herb's, can also be used. You only have to add 1 tablespoon of the Green Ban per quart of water. For Outside: Jungle Rain's an excellent soap based product which has citrus oils and peppermint soap. See also ant barriers that you can spray, later in this chapter.

If you need to use something fast then I would suggest that you use vinegar. Yes, plain vinegar will do. Use it straight.

To Sum Up: A big step in controlling ants is understanding how ants use trails to find their way around. Once you have started to keep the ants out of your house you must continue to fool them into staying out of your house. The Ant Cafe is an important tool towards this end. We will cover The Ant Cafe more fully in step 3.

Stay on the ant trails: Once you have gotten the ants out of the house, you have to keep them out. Use Dr. Bronner's Peppermint Soap (1 tablespoon per quart water) to spray around the outside of the house. Spray where the ants trails are found, usually along the wall and ground. Ants will leave trails almost anywhere so look carefully for them. (Avoid spraying directly on plants as this mixture is too strong and it will burn the plants.) This mixture is strong enough to kill ants on contact and will provide protection for a day or two depending on the time of year and conditions of the ant colony.

STEP 3.....THE ANT CAFE™

The Ant Control Center

The Ant Control center is a simple finch bird box with a secured lid. It has a small hole in the front through which the ants enter. This hole must be small enough only for ants; ¼ inch is enough. The box can be bought already made for baby birds to feed from; it's called a finch box. It costs about $5 and is available at any pet shop. It's worth the cost. It will last you several years. Inside the ant cafe goes an eight-ounce plastic cup to hold the ant gels #1 or #2. Use the wood screws to make sure that the box lid is secured so that small children and animals won't be able to get at the ant food.

Put your sugar source (honey, white sugar, or corn syrup) and water into the Ant Cafe cup (see liquid ant control gel #1 below), screw the box lid down, any place where ants are seen and where the Ant Cafe can be secured down. This will usually be at four locations along the outside of the house. Cover the cafe with large rocks so only the ants can get at it. Keep it out of direct sunlight as the sugar water will ferment. The ants will enjoy this free meal and will consider it a food haven. Do this for the first two weeks or so, keeping the Ant Cafe cup filled with some kind of sugar water. It is important to establish the cafe as a food source. This alone may solve the problem and you won't have to use a poison (boric acid) to kill them.

Gradually, over a period of weeks, move the cafes away from the house. You will notice that the ants have stopped going into your house and are going into the Ant Cafe. At this point, you can change the mixture and instead of honey water, use instant grits or yeast. Just place ½ cup full inside the ant cafe. The Ants will take it back to the colony and eat it.

Once eaten, the grits or yeast will expand and kill the ants. Do this just to get the ant population down at first, then alternate between feeding them honey with boric acid and using the grits or yeast.

HERE ARE A FEW SAMPLE FORMULAS FOR THE MIXTURES YOU'LL USE INSIDE THE ANT CAFE: EXPERIMENT!

Liquid Ant Control Gel #1

◆ 7 oz. water

◆ any sugar source, honey, molasses, etc.

Use a different source of sugar each month. One month use honey, next month use white sugar cubes, the following month use corn syrup, etc. Ants are smart and learn very fast and they also remember what they have learned! Mix it up - for example, blend 2 tablespoons of honey in the 8 oz. plastic cup filled with water, and you can add a little white sugar to make the syrup sweeter. To prevent any fungus from growing in the mixture buy some source of sugar that has a preservative.

Liquid Ant Gel #2

This is a protein version of above. Instead of sugar, use protein sources like cat or dog food or peanut butter. In this case allow the mixture to over flow the cup so that the ants will find it. It is very important that you show the ants where this feeding station is located by placing pure honey outside the Ant Cafe as well as across their paths. Once their scouts have located this feeding station, they will always come here first. Secure the Ant Cafe near known ant trails around the house and in the garden, place under fruit trees (hidden under rocks), etc., always out of direct sunlight. Use regular honey and place a few drops directly outside the Ant Cafe, to let the ants know it's there.

The stations have to be refilled every month (or as often as needed). If fermentation occurs, wash cafe out with the hose and refill with new gel. For this reason you should make only enough ant gel for use at one time. A fungus may also grow from the mixture; this is normal—simply wash and replace. To avoid this problem just add a little bit of any sugar source that has a preservative in it. Remember to change the recipe every time you make it.

You can also place the Ant Cafe inside the house in places like the kitchen. Make sure you screw down the lid so that no kids, dogs, etc., can get at it. (See info on using boric acid in index.)

Hominy grits anyone?

Using any of the instant grits or grains on the market will work here. The key is expansion when wet so it is important to keep the product dry for as long as possible. I suggest replacing the 8 oz cup with a small bowl. Placed inside the ant cafe, it will stay dry for a day or two depending on the weather, and other water sources around it. If it gets wet just change it with a new batch. I would alternate this mixture with the sugar and protein mixtures.

STEP 4.....FEED EM!

This is the most important step: teaching the ants how to use the Ant Cafe and then making sure they keep going there. You will find that you

---

IMPORTANT HINT:

Every month change the source of sugar. One month use white sugar, the next month use molasses, etc. If they are not going after the sugar, try peanut butter instead. If you're having an imported fire ant problem see About Imported Fire Ants (Solenopsis Invicta). Remember, do not use any boric acid until two or three weeks (a month is better) have gone by and, before you do, see "using boric acid" in index.

It has been said that there are more ants on earth than there are stars in the visible sky - but who counts that high?

Citra Solv: This can be a good alternative spray source for you to

control ants inside and outside the house.

must set aside a small amount of time each week to check the ant cafes. They must be kept clean. The sugar mixture tends to go bad in hot weather and must be changed regularly.

Step 5.....Back to the Ant Barriers

There are many natural ant barriers which you can use to keep the ants from getting into places you don't want them to be in. Ants will know that you don't want them in certain areas by using a barrier that they will not cross or do not like to cross. They are not stupid and will follow the path of least resistance. But you have to keep at it.

Some Ant Barriers that you can spray:

Soap: As I have mentioned before, using a natural soap to keep the ants from coming into your house is an important step in natural ant control. Use any kind of natural soap if you can't find the soaps that I mention. The main point is to avoid using any chemical-based soaps that could cause you and your environment serious damage.

More on using soap sprays: Using your indoor spray mixture, start by spraying the kitchen area sand then the bathrooms. Try to work your way from the center of the house out towards the edges. Spray lightly in the areas where ants have been seen. Don't forget the bathroom.

You might tell the ants the day before you do this that you want to make a deal with them. Tell them that you are willing to feed them outside your house and that you don't want them in the house, that you are going to spray the soap to make sure they leave the house, and that you are going to confuse them by hiding their ant trails.

Oil: There are many natural cooking oils that you can use which, when added to the soap, will keep ants off your plants, etc. Always try to use a natural-based oil instead of a chemical- or petroleum-based oil. A hot oil, like Tabasco sauce or Sesimin Hot Oil or Dragon Fire Oil, or any type of hot cooking oil will work. Tree Tea Oil works great here, too.

DE: Garden grade diatomaceous earth (DE or Dia-earth) can be used as a dust or added to water as a spray. You can add 1 tablespoon DE per gallon water and then strain it.

Hot sauces: Learning to use various hot sauces for ant control is fun and exiting. Again, stick to natural ingredients instead of those with a lot of chemical additives. See the index for more info on using hot sauces. As with the DE, mix the hot sauce with water and then strain.

Citra Solv: As I have mentioned before, this can be a good alternative spray source for you to control ants inside and outside the house.

Compost tea: Compost tea when properly made will control ants and many other types of insects on your plants. A good blend is compost tea made with molasses and Jungle Rain.

To 5 gallons of water add the following:

- ◆ 1 cup compost in a panty hose tied into a tea bag.

- ◆ 1 cup molasses

- ◆ 1 cup Jungle Rain

Mix the molasses and the Jungle Rain into the water, stirring for at least 5 minutes in one direction then 5 minutes in the other. This will increase the oxygen/energy level of water. Then place tea bag so that it is suspended in the water. Allow this to sit for two days or until you can smell it. This will depend on the air temperature. Add this mixture to sprayer. Spray on plants being effected by the ants and any other creatures. Spray as often as needed, but avoid spraying on hot days.

Hot water: Here is an ageless method of removing unwanted ant colonies from your lawn and other areas. Just pour water as hot as you can get it down their ant holes - avoiding plants or burning yourself. To increase effectiveness you can add tobacco tea by previously making a gallon of hot water and adding 1/2 cigar to it. Allow water and cigar to sit over night then strain it into a 5 gallon container to which you will add the boiling hot water as above.

Some barriers that you use as a dust:

DE: Garden Grade DE makes an excellent temporary barrier. It will last for only a few days and is only cumulatively effective - that is to say that an ant might walk over it and not die right away! This is best used outside the house.

Compost Tea
Compost tea when properly made will control ants and many other types of insects on your plants. A good blend is compost tea made with molasses and Jungle Rain.

Pyrethrum: Pyrethrum is a natural product that can be used safely inside the house. You can dust certain areas that are being effected by the ants. (For more on pyrethrum, see later in this chapter.)

Boric acid: While boric acid is not generally recommended as a dust, it can be used in out-of-the-way places that the ants frequent or on rugs if it is brushed into the rug. Otherwise it is best used added to their food source. (For more on boric acid seethe index.)

More Ant Barriers: There are so many natural ant barriers which you can use to keep the ants from getting into places you don't want them to be in.

A safe ant barrier: Pyrethrum powder and diatomaceous earth make a great mixture for controlling ants. Use 1 lb. Dia-earth and 5 oz. pyrethrum. Mix well. Avoid breathing. Dust where ants are found. Use outside - around plants and on ant trails. Use inside - under cupboards, sprinkled around windows, in cracks. Powdered pyrethrum is available from ARBICO, REAL GOODS, GARDENERS SUPPLY to name a few. To increase effectiveness you can also add any type of hot chili powder. Pure pyrethrum is 100% safe to mammals and birds but deadly to everything in the garden, including beneficials, and to fish so avoid over use.

There are many products on the market which use liquid pyrethrum. Pyrethrum by itself is an excellent, safe pesticide to use against a variety of insects. Pure pyrethrum powder is best to use here as it does not have any additives. One of the problems with using liquid pyrethrum products is that a product called Piperonyl Butoxide is added as a booster. Do not breathe this or get it into your eyes; if you do, wash with water.

Making a pyrethrum paste: Pyrethrum powder does not blend well with water. You have to first add a small amount (a few drops) of soap to a cup of warm water and then add a tablespoon pyrethrum to dissolve. The strength should vary according to the insects you are dealing with. You will need to strain it to avoid clogging up the sprayer. The amount you'll

make is determined by the amount of liquid you'll want to make. A paste can last for a long time so make a batch and save for later use.

Read the ingredients very carefully since many pyrethrum products use the synthetic form of pyrethrum (pyrethrin) which is not the same!

Making your own Plant Guardian™

- ¼ lb.   Dia-earth (not pool grade) see index for more info
- ¼ lb.   rock dust
- ¼ lb.   kelp dust (or seaweed powder)
- ¼ lb.   organic alfalfa meal
- ¼ lb.   organic pyrethrum flowers, dust
- ¼ lb.   organic cayenne pepper or Texas Gunpowder Chili
- 5 oz.   sulfur dust (optional)

This formula contains diatomaceous earth, pyrethrum dust, kelp, rock dust, cayenne pepper or Texas gunpowder Chili (available from Mo Hotta Mo Betta), and sulfur dust. It can be used as a dust, painted around tree trunks, as a paste, or made into a liquid and sprayed. Mix all ingredients well. Avoid breathing in or contact with eyes. Dried peppermint can be substituted here for pyrethrum. Remember, the pyrethrum will kill, the peppermint will repel. Use only a small amount for best results. Load mixture into a Pest Pistol or Dustin Mizer to dust plants being attacked by pests, or you can make a spray by mixing 1 cup paste with enough water to dissolve, then strain into 1 gallon of water. Add 1-5 tablespoons of Tabasco sauce or Cyclone Cider or any hot sauce. You can also use various other 'HOT' mixes available on the market such as Texas Gun Powder (not the type you put in guns but the type you make a chili out of). This organic paste is effective against ants as well as snails, slugs, earwigs, spiders and a wide variety of bugs. It is 100% organic. Use the Pest Pistol to form a line around tree trunks or around plants that are being attacked by ants. Repeat as often as needed.

A NOTE ABOUT USING BORIC ACID

Sometimes it is not possible to control the ants by just feeding them. When the imbalance is great enough there is infestation. An infestation is indicative of a greater problem. Correct that, and you will be better able to correct the ant problem. However, you will have to reduce the ant population in order to re-gain partial balance. The addition of boric acid to the ant gel will do just that. The ants will find it and take it back to their colonies. It won't kill them all, but it will decimate their ranks. They will then start over, but at a lower level.

You must make the boric acid just strong enough to slowly weaken the colony. If it is too strong it will kill the scouts right away before they can make it back to the colony. Use 1 teaspoon per 8 oz cup to start with. See using Drax for more info.

Even though boric acid is also sold as an eyewash, it is dangerous when taken internally. So avoid cuts, and please be very careful using it. The mixture should be a thick fluid (like honey). You should make just enough for your immediate use. Boric acid can be bought from Arbico, Real Goods,

Gardener's Supply Company, and at your local nursery store.

Boric acid used in concentrated amounts is dangerous to bees, cats, dogs, other small creatures, trees, and humans, so be careful not to use it outside the Ant Cafe. Keep out of reach of children. Do not spill around plants or trees; if you do, water well.

If you suspect that your child or animal has gotten into the Ant Cafe, get to a doctor ASAP!! (The amount of boric acid used in this formula is deliberately very low to insure that if accidental ingestion should occur, the mixture would not be strong enough to kill). Nevertheless, do not leave any amount of gel sitting around. If ingested, drink milk and induce vomiting and contact your doctor at once! Always follow instructions on labels of products that you buy. Never pour ant gel into any type of drinking container, soda bottle or in anything that would confuse someone into thinking that it was safe to drink. Always label the ant mixture as poison and write: Keep out! To be safe avoid using if their are children at home or dogs, cats, other pets. Think about locking down Ant Cafes with screws to keep them from getting at it.

To Kill or Not to Kill

Is that the question? I forget the answer! Anyway, the idea of using boric acid only when situations require it, is that using any kind of poison is not such a good idea really. Kill only when necessary not out of habit or because its easier. Poisons are never the answer. What are you gonna do, kill off all the bugs? Not in this lifetime! So what's the answer? That is what this book is all about, isn't it? Always seek balance. The bugs are just telling you when there's an imbalance. Really!

About using DRAX or TERRO

Both DRAX and TERRO are made up of 5% boric acid with a sweet base for the inert. These store-bought formulas should be adjusted so that the level of boric acid that the ants get is strong enough to kill the queen, etc. but weak enough not to kill the workers who bring it to her. A good way to determine this is to first provide a 100% pure sugar source, then, after the proper length of time (two weeks to one month), slowly switch to boric acid by first adding 1 teaspoon boric acid into the formula, and noticing if the ants are dying at the site and consequently have left the mixture alone; if so, it is too strong. If the formula for the ant gel is killing too many ants too fast and the ants are no longer going there, then use ½ the formula for boric acid (1/2 teaspoon). The Ant Cafes are a very important tool in your ant control program. They are useful for helping to control many other insects as well.

If, after a month, the ants are not dying, then increase to two teaspoons per cup. Again, watch the ants for a few days and note activity. If the ants stop coming and/or die while at the cup, the mixture is too strong. Remember, please use the boric acid only as a last resort and don't use it all the time. I suggest getting 4 Ant Cafes per home and nailing or other wise securing them outside, around the house; 4 per garden, nailed to raised beds, and placed around fruit trees being bothered by ants. Ant Cafes will provide continuous control of ants, carpenter ants, earwigs, and roaches. Many other insects are also attracted to this mixture. You can make your own ant gel or you can buy DRAX or Terro from Real Goods, Gardeners Supply Company, Peaceful Valley, or Harmony Farm Supply. These gels work fine for one or two

A NOTE ABOUT THE WARNING LABEL ON THIS PRODUCT
The EPA classifies pesticides into four toxicity categories but they use only three "human-hazard" warning signal words for pesticide labels: Danger (Poison), Warning, and Caution. Therefore natural substances such as bacillus thuringiensis (BT), garlic, Safer Soap, pyrethrum, Dia-earth, etc., carry the same signal word as Sevin, Trimec, Dursban, Diazinon, and many other chemicals. The natural substances should not be categorized with these chemicals, yet the EPA has not created a fourth signal word. Until this happens, the consumer, if not informed, will pick up Sevin, etc., thinking that these products are as safe as BT, etc. He or she thinks they have only to use caution. This however is not the case. Don't be fooled into believing that chemicals are as safe as natural products. Conversely, even Organics must be used with care, since many Organics such as tobacco can be dangerous if misused.

The Invisible

Gardener

says:

"Before you

add boric

acid, ask

yourself if it

is necessary to

kill them or

was the sugar

alone keeping

them out of

the house?"

months until the ants get wise to them and leave them alone. They are too strong and will kill them too fast not allowing the boric to get back to the colony. That is why it is best to blend with your homemade gel.

STEP 6.... DOING THE WORK ~ DANCES WITH ANTS

Learning how to make the Ant Cafe an active food source for the ants is a very important part of the treatment. I cannot overstate this. The ants will not at first be able to find the new food source—for a variety of reasons. Sometimes they will not want it and ignore it. You must make them understand that this is a food source for them. This is what we call "Dances with Ants". Therefore it is good to go slowly when first introducing the Ant Cafes to the ants.

Providing food and water for the ants outside while using the soap spray inside may be enough to control your ant problem inside the house. We call this "Dancing with the Ants" also, because it takes time to get the ants to go for the Ant Cafes rather than into your house or on your roses.

Concerning the container used to build your ant cafe: I have experimented with many different types of containers in order to find the one which worked best. There are many problems with using plastic containers due to the pollutants associated with their manufacturing process (damage to our OZONE layer, damage to our oceans, etc.) so I do not recommend using plastic unless it is made properly. After trying out many different types of containers and after making different types of my own, I have decided that it is easier to go to your local pet store and buy a finch bird house! So the Ant Cafe is modeled after this bird house shown here.

Here are some important points to keep in mind when you are looking for that "perfect" Ant Cafe: It is very important that the unit can be locked, screwed down or in some way made impossible for children, dogs, cats, etc., to get into. We are only after ants here! If you are concerned about children getting into it then don't use any boric acid—use only sugar, honey etc.

The container must hold enough ant gel (or plain sugar water) to last a week or two and be small enough to be hidden from view.

Continue to spray the peppermint soap as often as needed inside and outside the house while checking the Ant Cafes at least once per week to make sure the ants are using them; clean them as necessary. As I mentioned before, a fungus may grow in the sugar mixture since ants carry their own fungi on their bodies. Just clean the cafe out by disposing of the used liquid in the trash and rinsing the cup. Replace with a clean solution. Spray something different every month. See below.

ORGANIC ANT CONTROL SPRAYS

Here is a list of some formulas for making organic sprays, what they are, how to use them, and some sources.

To recap again, any organic biodegradable soap, or scented mixtures such as mint extract, garlic oil extracts, Tree Tea Oil, or Tabasco sauce, etc., can be used to spray on ants. You can also use scented soaps such as Dr. Bronner's Peppermint Soap, Green Ban or any natural concentrate made without chemical additives such as Jungle Rain, etc. Solar Tea made from any of the mint family such as peppermint may also be used. (This is tea made by placing the tea bags inside a glass container filled with clean water and leaving it in the sun for 24 hours. See making solar tea in appendix.)

A Safe Ant Spray for the Garden

1 gallon Sprayer

2 tablespoons any natural biodegradable soap such as Dr. Bronners Peppermint Soap

1 tablespoon Louisiana Tabasco Sauce

5 drops sesame seed oil

5 tablespoons Jungle Rain

Dr. Bronner

Add Dr. Bronners Peppermint Soap, sesame oil, Jungle Rain and Tabasco sauce to gallon water. This is to be sprayed directly on ants. Please note that instead of Louisiana Tabasco Sauce (which is organic and has no additives), there are several hotter products on the market. The hotter the better, which reminds me about a catalog called "Mo Hotta Mo Betta", a great source of various hot chilies and other hot sauces.

The Ant Spray Formula for Inside: Add 1 capful (tablespoon) of biodegradable soap per quart of water. You can also add 2 drops mint (if you like the smell). Always check the ingredients to be sure that no chemical additives are in it. It is almost impossible to find pure foods so go to a good health food store to buy your materials, or make your own. Try buying your materials from mail order suppliers such as ARBICO or NITRON, REAL GOODS, Gardeners Supply and many other mail order sources.

You will need a good quart sprayer. You may want to experiment with the mixture to see what works best. Use this spray as often as needed; usually once per week will do.

Spray around the outside of the house and any places where you see ant trails or where they are entering the house. Spray only on the ants. Flood the colonies with it (usually under rocks) and their traveled paths, around the base of your house, and other entrances, but DO NOT SPRAY DIRECTLY ON PLANTS AS THIS MIXTURE WILL BURN THEM! If this happens, wash off with water. You can use just the soap and water when spraying on walls.

Citronella Oil: Citronella oil comes from citronella grass and is an excellent oil to use for many purposes. Try smelling it first to see if you like the scent (not everyone does). You can use it alone or you can add a few drops per gallon to the above formula to increase effectiveness.

Tea Tree Oil: Tea Tree Oil also a great addition here. Ants do not like this oil as it interferes with their markings as well as interferes with the signals which they transmit and receive through their antennas. This is an excellent oil that you can use inside and outside to control ants with. Try your local drug store for this.

You may also try other essential oils.

Liquid Plant Guardian Formula: The same formula as above is followed except that ¼ lb. of pyrethrum paste is added to 1 gallon of warm

For Immediate Relief Inside: To control ants immediately use Dr. Bronner's Peppermint Soap. Add 5 drops Dr. Bronners to a quart water. Spray directly on ants in the house. Start in the bathroom and work your way towards the outside of the house. This will last for only a few days.

water slowly, while stirring. Allow to sit for one hour and cool. Then strain this mixture through a fine cloth. Use 5 to 20 drops per gallon for most bugs. You will have to experiment and find the best amounts to use for your own areas. Use either citronella oil (at 5 drops per gallon) or, if you don't like citronella oil, you can use Tabasco sauce or any hot sauce (1-5 tablespoons per gallon) or Tea Tree Oil. You can also use Cyclone Cider at 5 drops per gallon. WOW! These will control, kill, or repel a wide variety of bugs on contact.

A NOTE ABOUT FIRE ANTS

The imported fire ant IFA (solenopsisinvicta) is believed to have come into the USA sometime around 1930 through a port of Mobile, Alabama. This species of ant is very different from its American cousin. IFAs are much more aggressive and have the ability to sting many times with a toxic venom. These little guys will not take a no for an answer and will attack you if you are in their way or if you have stepped on their mounds. The IFA also build bigger, more powerful colonies, with an extensive underground system of tunnels that can reach 50-100 feet in any direction from their main mounds, which have been known to be as tall as three feet! These guys, though originally native only to Argentina, had claimed 27 millions acres by 1957, and 230 million acres since 1985. In 1995 they began to move into California and Arizona. They have attacked and taken over territories normally claimed by their American counter parts.

When they first arrived they were given help in the form of the US government's efforts to eradicate them. The government funded widespread spraying of the pesticides heptachlor and mirex during the 1950s and 1960s. Guess what - while the spraying killed off millions of the ants, they also killed off any competition allowing the remaining IFAs to expand at will. The spraying were eventually stopped but not until it was too late and the IFAs were here to stay. They quickly spread into eight southern states, into Oregon, and Washington, then, more recently, into Arizona and now into California.

Millions of people are stung every year by IFAs making them very dangerous to those people that are allergic to their bite. At least 50 people die each year due to allergic reactions to the IFAs bite. This can be prevented with immunotherapy using whole-body extracts of the IFAs. Since up to 60% of any urban population is bitten by IFAs, this pre cautions a very important step to take. If you are one of these folks who are allergic to the IFA's bite then see your doctor. IFAs have been known to attack people in their homes, gardens, and almost any place where humans and ants meet. This is a very different case than living together with harmless ants. To make matters worse (if they can get worse), these ants no longer fight amongst themselves for food or territories, instead they have gone from a single queen per colony (which fought other colonies) to multiple-queen colonies which have lost the ability to distinguish between themselves and the ants of other colonies, so they live together and share in the food gathering, etc.

Where a single queen colony would contain 200,000 ants per colony, with about 50 colonies per acre (in urban areas, as much as 20 colonies would have fought over a single ¼ acre lot!), a multiple-queen colony can have as many as 400 multiple-queen colonies per acre (or as many as 100 mul-

The Invisible Gardener says: "Something to remember is that you really don't want to kill everything in sight. The idea is to learn to live together in harmony and killing pests should only be done as a last resort."

tiple-queen colonies per ¼ lot!) living together in peaceful coexistence! With this type of multiple-queen set up, IFAs can withstand greater cold and are able to come back faster from pesticide attacks by humans, with very few animals or other creatures being able to withstand their onslaught.

Another problem is that the red imported fire ant (IFA) and the black imported fire ant (solenopsis richteri) are interbreeding, creating an even stronger hybrid which is more tolerant of pesticide attacks and more resistant to cold (they can live farther north than before). One of the fire ants' natural enemies, the South American Phorid Fly, is being considered as a possible control alternative but if lessons in the past have been learned more study is needed before this fly is imported into the USA.

The ant control methods described above will work against the IFAs just as well as against any other types of ants. Boric acid will kill them just as surely as it will kill off other ants. Just remember that the strength of the boric acid must be enough to kill them but not so strong that the ants cannot take this mixture back to their colonies and eventually to their queens.

IMMEDIATE HELP WITH ANTS

For Immediate Relief Inside: To control ants immediately use, Dr. Bronners Peppermint Soap. Add 5 drops Dr. Bronners to a quart water. Spray directly on ants in the house. Start in the bathroom and work your way towards the outside of the house. This will last for only a few days. You will have to spray daily in bad areas. Because it's a two-way street we must try to understand why ants do what they do. While ants learn quickly, people do not. So don't be ashamed that it will take you and the ants some time to understand what you want of them and what they want of you. They are usually after food and water. So clean up after yourself. Don't leave sugars and other foods around for them. Sometimes as simple a thing as cleaning up inside will be enough to keep the ants outside.

Immediate Relief Outside: Learn to use various soaps as a natural barrier to keep the ants away and out of your house. The many types of soap available at your local store can be used directly on the ants as an immediate deterrent but you would be better off if you used a soap made from natural ingredients. Preferences of mine are Dr. Bronners Peppermint Soap, Jungle Rain, and Citra Solve, to name a few. These can be mixed with water and sprayed directly onto the ants and the ant trail. The amount that you mix with water depends on what works. So try out a small amount on the ants first and see what happens. If the ants don't die then make it stronger. The Jungle Rain should be used only outside as the orange oil is not good indoors but DR Bronners or Citra Solve are great indoors! Use what smells good to you. Spray the soap around the outside of the house as well as inside to keep the ants out.

A few things you must understand...

A few things to remember in controlling ants:

1...Have patience. Controlling ants takes time, don't be in a hurry. They have been around for a long time and there are many of them!

2...Keep doing it. Don't give up. Keep doing the steps. It will work.

3...Don't give in to the chemicals. They will only make it harder for the natural system to work.

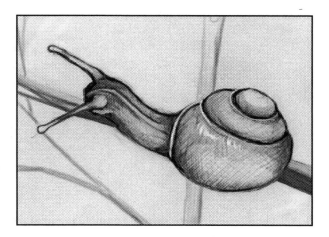

2

Snail Tales

How to Control Snails Organically

Most problems with snails can be explained by the law of cause and effect. The snail problem that many gardeners encounter is an effect - a reaction - and not a cause. Snail infestation occurs when the soil is no longer alive because the organisms within it have been killed with chemical fertilizers, pesticides, or thru otherwise improper soil care. And the continued use of snail killer year after year further upsets the balance and destroys the soil. This is the cause. When the soil is dead, the delicate balance which nature maintains has been upset. Deal with the cause and the effect - in this case the snails - will disappear.

Here are some steps you can take to control snails on your property.

Step 1: Dealing with Snails

Whenever there is an infestation of any one pest, the balance of nature has been upset. Anything not naturally found in nature contributes to this imbalance, including chemical fertilizers, pesticides, herbicides, weed killers, snail killers or any other poison and/or synthetic toxin. If you have been using any form of chemical snail control year after year to keep snails away from your flowers, vegetables, etc., the first and most important step is to stop using any form of chemical snail bait whatsoever.

Another reason to stop using snail poison is that it also kills birds and other friendly critters that you may like having around your home[1]. If you want to deal with snails effectively, you will succeed only if you understand that chemical fertilizers are as bad for your soil's health as pesticides are bad for your health. You must be willing to commit yourself to growing without chemicals of any kind. Because your plants need a period of transition, start by reducing your chemical use by half during the first 3 months, then completely stop the chemical use after that period.

Step 2: Withdrawal

Allow yourself and your property time to go through this withdrawal period[2]. This is just as important to you as to your property. Chemicals only cause imbalance. The withdrawal period is a critical time for you and your property. Without the snail bait, the snails will seem to increase at first[3], you must therefore employ <u>commando</u> tactics.

---

[1]Dogs have been known to eat this stuff and die from it!

[2]That is because they will be increasing at first! Don't worry!

[3]A good withdrawal period is one year. This will provide you enough time to get used to dealing with the snails without the chemicals and the earth enough time to recover gently.

Step 3: Handpicking

Handpicking snails and slugs[4] is one of the fastest and safest ways to reduce their population. Snails are best picked after dark[5] or on cloudy days when they are most active. Water well a few hours before picking to promote activity on their part. Throw a snail picking party! Tell your kids that you will pay them something per snail picked. A good price might be a penny per snail (100 snails equals 1 dollar). Have plenty of flash lights available. Provide gloves. Have buckets filled with salt and soapy water, diatomaceous earth (DE), salt, or rock dust layer. Throw picked snails into a bucket for disposal later.

Alternatively, you can crush the snails and place them into the ant cafe as ants love to eat snails. If using rock dust, crushed snails can be added to compost pile. If you have chickens, feed the snails to them. Pick up every single snail you see and destroy it. Snails can produce over 300 eggs per day during laying season and the eggs can stay buried in the ground for up to 11 years and emerge when the time is right for their survival. Another method is to attract and trap them. (See Snail Cafe).

> *"The day you stop using chemicals is the day you start to regain the eco-balance of your property."*

Step 4: Restoring the Balance

Increasing the energy level of the soil increases the health of the soil. This process naturally reduces pests and diseases. The best way to increase the energy level of the soil is through proper organic nutrition. Compost that is rich in minerals and bacteria naturally increases the energy level of the soil. The addition of rock dust further increases it. Higher energy levels support a more beneficial diversity of life.

Step 5: Natural Snail Control Methods

Attracting and trapping snails is one of the most effective natural snail controls. The best way to do this that I have found is through use of the Snail Cafe.

In my search for a more effective way to deal with snails, I found that people generally use one chemical or another to rid themselves of snails. They continue dumping chemicals into the soil even though their snail problem persists, when what they should do is concentrate on nutrition and healthy soil, instead. And so you have to use a method other than anti-snail chemicals to get rid of the existing snail population.

During the transition period when you are creating a healthier environment around your home, a good way to get rid of the snails is through the use of a Snail Cafe.

> *"Beneficial diversity is described as the various inter-relationships between living organisms and their contributions to the whole."*

---

[4]And especially picking their eggs! The eggs are found below ground in clusters or under rocks and boards. By cultivation of the soil you can expose the eggs and simply remove them by throwing them away in a plastic baggy.

[5]The best time is really at night since they do their work at night.

THE SNAIL CAFE™

The cafe should be made from a natural product such as wood. Wooden bird houses make excellent Snail Cafes. Go to any bird or pet shop and look at their bird houses. A good unit must be big enough to both hold a bowl and to allow snails to enter easily. The cost to build a Snail Cafe is minimal. Pick a bird house that allows enough room for the snails to enter, and one that has a lid that can be opened and closed. This prevents the snail bait or snail brew from getting into the soil and insures that you will only kill snails (if using boric acid). This unit will also protect the snail bait/brew from rain or sprinklers. Inside of the cafe you place your homemade organic snail bait or snail beer. You should alternate between using snail bait and snail beer. This will increase the effectiveness of your Snail Cafe. Place in a shady spot. Remove snails everyday, weekly or as often as needed. Discard the snails into the compost pile or use in the Ant Cafe[6]. Replace with new snail bait or snail brew. Use as many Snail Cafes as needed[7].

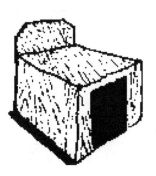

The cost to build a Snail Cafe is minimal. Pick a bird house that allows enough room for the snails to enter, and one that has a lid that can be opened and closed.

THE SNAIL BAIT
- 1 lb cornmeal and 1 lb instant hominy grits
- 1/4 lb flour
- 1 oz salt or baking soda
- 10 oz DE or rock dust
- 1 tablespoon brewers yeast

Mix all ingredients together and place into snail hotel. Feeding the snails with this mixture will dehydrate them. The instant grits will then expand and kill them, or, if they crawl over it, it will dry them up. The bait must be kept dry and replaced when wet. Be patient, as this process will take time.

Boric acid can be used (one tablespoon in the above mixture) but only as a last resort and if the snail infestation is overwhelming.

"Snails can produce over 300 eggs per day during laying season and the eggs can stay buried in the ground for up to 11 years and emerge when the time is right for their survival."

THE SNAIL BREW (TO ATTRACT THEM BY THE THOUSANDS!)
- 1 gallon plastic bottle
- 1 tablespoon brewers yeast
- 1 quart apple cider or Vinegar
- 1 quart cheap wine or beer
- 1 tablespoon natural soap like Jungle Rain
- 1 cup warm sugared water or honey.

In a one-gallon bottle, combine one quart cheap wine or beer[8] and one quart apple cider or vinegar. Mix in one tablespoon of brewers yeast and enough warm honey-water to fill. Add one table-

---

[6]After crushing!

[7]It is important to keep the cafe out of direct sun and to keep the snail bait dry.

[8]Many have heard about snails and beer. When I was in the sixth grade I wanted to know if this was true. I went out and with the help of an older friend bought 10 different types of beer plus a few bottles of wine and a few other things such as vinegar, wiskey, brandy, etc., and yeast. The snails preferred the stronger drinks such as the whiskey and brandy, but the best results came with the use of apple cider, vinegar, and yeast. Whenever I added yeast to any of the strong drinks, the snails went nuts for it! Beer does work but will fizzle out and must be replaced every day to be effective while the stonger drinks last for a week before having to be changed.

"Providing snails with hiding places makes them easier to locate and dispose of. Old nursery containers placed on their sides make excellent temporary hiding places for snails."

spoon of any natural soap such as Dr. Bronners Peppermint Soap, Shaklees Basic H, Jungle Rain, or Citrus Soap (see natural soap chart). Do not shake, but stir well. Pour into plastic cup or bowl and place into the cafe[9]. Make sure the lid is open to allow the snails to enter. Place near their favorite hiding places. You will have to check it regularly, to remove any dead snails and/or replace with new snail brew. This works best at night.

Providing snails with hiding places makes them easier to locate and dispose of. Old nursery containers placed on their sides make excellent temporary hiding places for snails. You can place a bowl of snail brew inside these containers to attract them. You can also use boards laid on a rock to allow snails to crawl under. Pick them off regularly. This provides day time shelter for them so look under it during the day.

SHOOT EM UP

If you do not want to hand-pick the snails you can always try the Cowboy Method. Prepare your ammunition by mixing one-part salt to one-part cayenne pepper in a salt shaker. Carry the shaker with you and sprinkle the contents on any snails or slugs you encounter. This causes them to dehydrate and die[10]. Another method is to use a spray bottle of water with one tablespoon ammonia and two tablespoons of any natural soap in it. If you don't like ammonia, you can mix two tablespoons vinegar, two tablespoons soap, and two tablespoons tabasco sauce. Spray either of these two mixtures on the snails to kill them. A third mixture is to use Tabasco sauce (about 10 drops per quart water) and spray directly on the snails. Try also using just two tablespoons Dr. Bronners Peppermint Soap[11] sprayed directly on snails.

SNAIL BARRIERS

Silica sand, alfalfa me, rock dust, flour, cayenne (see also physical barriers in chapter 5), DE, wood ashes[12] and seaweed all make excellent barriers to keep snails from crossing over into areas out of which you want to keep them. Use as little as needed around plants to protect them from snails. Tangelfoot, made of castor oil and wax, placed around the trunk of a tree also makes an excellent barrier. Add cayenne pepper (or any hot barriers such as chili) to increase effectiveness. A foot wide barrier of any of the above should be placed around the edge of your garden or beds. Replace regularly. This will keep snails and other crawling insects from visiting your garden. Ocean sand works well, too[13]. Other barriers you can use are crushed egg shells or aluminum screening around raised beds—bend the edges out to keep the snails from crawling over. Hardware cloth tacked to the raised beds will also work.

---

[9]Using the apple cider, wine, vinegar, or beer to control snails works because of their high fermentation. Any fermented products will attract the snails—and many other insects as well.

[10]Be careful using salt as it will cause soil and plant damage if too much falls on the ground. Dust only the snails here!

[11]Any natural soap will work here.

[12]Go lightly with the ashes as it will raise the ph level of the soil.

DIATOMACEOUS EARTH *(DE OR DIA-EARTH)*

Diatoms (Diatomite) are either lacy-like snowflakes in appearance, or tubes, which are better. They are microscopic in size. Their color is usually somewhere between white and brown. They are found both at the bottom of the ocean and in fresh water lakes. Their by-product is oxygen. Three-quarters of the air we breathe comes from diatoms[14]. As the diatom dies, the dia-earth[15] falls to the ocean bottom or lake in which they lived. Layers of this silica form and do not degrade. After 150 millions years of this, we now have rich deposits of this substance. DE is chemically identical to quartz and white ocean sand. DE is totally harmless to mammals if eaten and is recommended to be given to your animals as part of their diet. Many years ago when tests where run on the toxicity of DE, it was found to be beneficial to the animals fed with it as they gained weight and seemed healthier then those not fed DE. Animals fed DE have fewer problems with intestinal parasites. Dia-earth is available from **Nitron, Arbico, Gardeners Supply, Peaceful Valley** and **Harmony Farm Supply** to name a few.

DE can be fed to animals with their food, or in a bucket mixed with a little grain. For cats, use one teaspoon mixed along with their wet food, once per week. For dogs, add two tablespoonfuls per week mixed well with their wet food. For horses, use one-half cup per bucket of sweet oats, or add to a bucket of drinking water.

HOW DIATOMACEOUS EARTH WORKS

Dia-earth (DE) is a dust that bugs attract by way of the static electricity produced when they move about. Nothing "bugs" an insect more then being dirty, hence cleaning themselves is a major part of their daily activity. DE clogs their breathing 'holes' (insects breathe through their outer exo skeleton), and removes oils that insects naturally produce to help protect themselves, leading to dehydration and death. This happens slowly, over a period of time that varies depending on which insects are affected, and on various environmental conditions. Usually a small amount will do. Often farmers use a lot of DE as they want faster results and are impatient.

DE can be blended with water and part soap and sprayed on the plants or insects. When it dries, it leaves a thin coat of DE behind. Use a little bit. Remember that it will kill beneficials as well so apply only where needed.

---

[13]Very salty so go lightly. Do not put directly under plants. It's okay for barrier around raised beds though.

[14]They carry on photosynthesis.

[15]Dia-earth is 90 percent pure silica.

Avoid Breathing DE

Breathing DE should be avoided because it is rather harsh on your lungs, like any irritant. If you try breathing flour, you will get the same results. So it is best to keep the dust down to a minimum. DE is not harmful to your health if used properly, but do be careful when using it. Keep it away from your eyes. DE is like millions of microscopic razor blades, and should be washed out with water immediately. Avoid rubbing your eyes, for obvious reasons. For extra protection, wear a face mask when using it. You can use Duster Mizer to apply (available at most mail order companies such as Nitron, Arbico, Gardeners Supply), or you can use a flour sifter or pest pistol.

DE prevents snails from laying their eggs and kills them by slow dehydration. It will significantly reduce the snail population over several years. Remember, the best results may take longer but are more permanent. DE should be dusted onto plants and placed as a barrier around them. Do not water for 24 hours after, to allow the DE to take effect.

An Organic Snail Spray

- 10 drops SuperSeaweed[16] per gallon
- 1/2 cup Diatomaceous Earth per gallon
- 1 cup Nitron A-35 and 1 cup AgriGro per gallon
- 5 tablespoonful bio-degradable soap, like Jungle Rain per gallon
- 5 tablespoonful tabasco sauce or Cyclone Cider per gallon

To 1/2 cup DE, add 5 tablespoonsful tabasco sauce or Cyclone Cider, add 5 tablespoonful bio-degradable soap such as Jungle Rain, Amway's LOC or Dr. Bronners's Peppermint soap. Slowly add enough water to form a slurry. Pour this mixture through a strainer into your gallon sprayer. Add 10 drops Superseaweed per gallon or you can use any natural seaweed concentrate available in your area or through mail order. Add Nitron and AgriGro. Spray this mixture once per week (during infestation) in vegetable gardens, around lawns, on fruit trees, etc. Spray plants and surrounding areas. For best results spray in the late afternoon.

An even simpler snail spray is compost tea!

See the Resource Directory for more listings; also check the chapter on making your own organic sprays.

Always test plants and crops to be sprayed to check for adverse results before spraying larger areas! If your snail infestation is heavy, double the DE and Nitron and spray more often. Be careful with spraying delicate flowers, and leaves of plants. It is best to spray under plants and trees (on tree trunks). Do not water for 24 hours following the treatment to insure maximum effect.

Organic Alfalfa Meal

Alfalfa meal is high in nitrogen and other natural bacteria. It makes an excellent snail barrier while feeding your plants at the same time. Sprinkle around the base of the plant; do not water for 24 hours following treatment.

Kelp

The use of kelp or any dried seaweed in your plan to control snails is highly recommended. Kelp will provide the soil with a great many trace minerals, all essential in maintaining a balanced soil. Kelp will also reduce the snail population by raising the salt level too high for them, while keeping it to a level tolerated by the plants and soil. Use various different types of seaweed. Different parts of the worldproduce seaweed that is rich in different trace minerals, and various seaweed varieties will complement each other.

See the chapter on making your own organic sprays for more information on seaweed.

---

[16]Alternatively, one cup of a natural liquid seaweed will also work here. I like Acadie or Maxi Crop.

SuperSeaweed

Superseaweed (or your own liquid seaweed blend) will be useful here as it is a blend of seaweeds from around the world. Spray the soil and plants with the Superseaweed once or twice a week to make certain that you have all the trace minerals that your soil and plants need. If you live near the ocean you can collect your own seaweed, but you must allow it to dry properly, separating the salt for later use. You should test the seaweed as well as the salt for toxins, because the ocean is getting more polluted. Also ask for test results on any seaweed product on the market today. See the Resource Directory for some excellent sources of seaweed. Superseaweed is available from The Invisible Gardener. (See Appendix.)

Mulching

Mulch is a highly recommended form of snail control. Its effectiveness depends upon the type of mulch you are using. I recommend alternating between pine needles, kelp, rock dust, various types of leaves, compost (makes a good mulch by itself if made with bark chips), greensand (too expensive as a mulch but I mention it here because you should add a thin layer once per year), flour (yes, regular flour makes a great mulch barrier!), bone-blood meal (a thin layer once per year will do it), and any other assortment of available mulches. The more varied the mulch the greater the success. Look around in your area for free mulch. The easiest free mulch available is horse manure which is at least 9 months old - when it's that old it has no smell at all. This makes a great mulch provided you also use assorted organic fertilizers and plenty of peat moss or pine needles to keep the PH down to around 6.5 (depending on what you are growing). Just spray the mulch with a natural soap to control the snails.

Making your own Sticky Stuff
- 1 cup petroleum jelly
- 10 oz. castor oil
- 1 oz. cayenne pepper
- 1/2 oz. tabasco sauce or other hot sauce

You can make your own Sticky Stuff by using petroleum jelly and castor oil blended together. Simply apply the resultant gunk to trunk of plant. You can also add 1 to 2 oz. cayenne pepper or another hot barrier and/or 1/2 oz. of tabasco sauce to this mixture to increase its effectiveness.

If you use Tangelfoot, you can make it work better by mixing in cayenne pepper or another hot barrier[17].

---

[17]Be careful using Tangelfoot as it can cause damage to the trunk of the tree. It's best to use duct tape first!

IG says:
There are 4 important things to remember about controlling snails:

1...Give yourself lots of time

2...Maintain healthy environment

3...Avoid chemicals at all costs!

4...Plant the proper plantings for the season and location where you live.

Edible European snail,
Helix

Natural Snail Predators

There are many natural snail predators. Birds love to eat snail and they are a very important part of their diet. Beetles prey on snails, slugs and other insects. Many snakes and lizards love snails and slugs. Toads are especially fond of snails and slugs and a pond can be a good tool for attracting toads to your area.

Snail predators are available from:
ARBICO
Peaceful Valley Farm Supply
Harmony Farm Supply
Rincon - Vitova

FLOUR OR CLAY

Flour or clay can also be used as a barrier. Using one cup flour or clay, add enough water to make a slurry, then add five tablespoons of cayenne pepper or another hot barrier, and stir well. This can be painted on tree trunks, or used as a dust around plants[18] .

A BIO-DYNAMIC SNAIL BARRIER

Old time bio-dynamic farmers used aged horse manure mixed with clay and seaweed and painted it as a barrier around trees. To do this use enough old horse manure to fill a bucket. Add one or two lbs of clay or flour, 1/4 lb cayenne pepper (or another hot barrier) and two lbs powdered seaweed to the manure. Then add enough water to make it into a paste and paint it onto tree trunks.

COPPER

Copper bands wrapped around the trunk of a tree or plant also comprise a good, non-toxic form of snail and slug control. Copper clips and barriers can be used in your garden by fastening them to raised beds[19] . There are many companies which sell these strips for your garden use, or you can make your own from throw away parts.

DECOLLATE SNAILS

Decollate snails are a very effective natural predator against the brown garden snail. This snail is smaller and has a shell that looks like a seashell. This snail prefers decayed vegetation and small snails are its favorite diet. (Larger snails will have to be hand picked.) When using decollate snails be aware that any barriers or sprays you use will effect the decollate snails, too. Use decollate snails only as a last resort[20].

IPM SNAIL CONTROL DUST

- 1 lb rock dust
- 1 lb dia-earth
- 1 lb kelp meal or seaweed powder
- 1 lb organic alfalfa meal dust
- 1 lb hot chili

Blend together. Use this as a dust around base of plants. See chapter on rock dust for more information. While DE is good for animals, it is hazardous to snails. When handling, avoid breathing it or getting it into your eyes as it is rather harsh. If you do get it in your eyes, wash them out immediately with copious amounts of water.

---

[18]To use as a dust don't add the water!

[19]Copper has long been used to control snails and slugsslugs. Some believe that the copper provides a slight electric charge that keeps them off while others believe that the copper itself is toxic to them. I use copper sulfate at one drop per gallon to provide control against snails and slugs. Use only in small amounts as the sulfate can damage plants as well as insect life.

[20] Check with your state for laws concerning these predators before ordering!

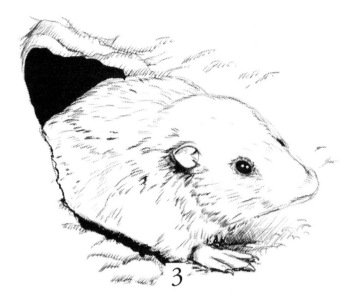

3

Gopher, Gopher

How to Control Gophers & Moles Organically

Pocket Gophers, or Ground Squirrels as they are sometimes called, can be as big a problem as ants[1]. Gophers are like little bulldozers. They dig a network of tunnels usually six to twelve inches below the surface[2]. Tunnels near the surface are for gathering food and act as highways to get around; deeper ones are for sleeping, raising the young, and storing food. Well-suited to burrowing, gophers have small eyes and ears that don't clog with dirt, and sensitive whiskers and tails to guide them in the dark. Gophers have long teeth for biting off chunks of hard soil or roots. They are sensitive to sound, and a variety of smells, as well as to light[3]. To successfully control gophers and other rodents, one needs to understand their senses of smell and hearing.

Gophers eat plant roots as well as bulbs. If you see a plant being dragged under ground you know you have gophers. They will eat almost anything including vegetables such as tomatoes, carrots, squash, lettuces, etc. They also love roses (the roots). They'll nibble on the roots of your fruit trees, the roots of most flowers, and tulip bulbs.

Gophers can do a great deal of damage[4] but are an important part of the natural ecosystem. I do not recommend the use of traps as they are inhumane and very cruel[5]. It is important that you properly feed and care for your plants as gophers are less likely to attack healthy plants. By top-dressing your lawn and property with year-old horse manure, you will not only be feeding your soil,

but also repelling gophers naturally as they do not like manure. You will also be encouraging natural predators as compost promotes healthy environments which always attracts the variety of life needed for there to be a balance.

Moles are insectivores; they eat earthworms, bugs, larvae, grubs, spiders, and centipedes and only occasionally nibble on greens and roots. Most of the time, they do very little damage other then through their tunneling, which can unearth plants and expose root systems.

How can you tell whether you have a gopher or a mole? Gopher mounds have dirt pushed off to one side of the mound while a mole mound will be volcano shaped with the center at the entrance to the tunnel (usually plugged up). Poisons do not work on moles! They must be trapped and moved away.

---

[1] And as easy to control!

[2] The deeper the tunnel, the older the gopher that dug it!

[3] This is the key to their control!

[4] It is far better to get used to them than to try and kill them off. A lot depends on where you live. In my garden, gophers come and go as they please. They are welcome as they do a lot of soil-turning for me! It is not hard to train them to only be in certain areas of your garden and to leave other areas alone. I like gophers coming through certain parts of my garden at certain times—they loosen the soil for me, eat grubs and insects, and then leave the garden for me and my vegetables!

[5] Unless you use the humane traps which don't hurt them and then move them to another area.

The Process

*PLEASE NOTE

Gopher Getter Midget and Jr. comes with its own poison bait—**tell them that you only want the empty container and not the poison inside.** This poison bait is made from milo grain and strychnine, a natural highly toxic poison derived from Nox Vomica of the nightshade family. This is a most poisonous material so if you do use it, **EXTREME CAUTION IS URGED**. It is natural, but highly poisonous. Keep the warnings in mind so as not to unintentionally poison any birds, rodents, animals or people. It must be placed in the gopher run, away from where birds and others can get at it. (One grain can kill a bird!) **READ THE DIRECTIONS. HANDLE AND STORE IT WITH CARE.** It also contains 35% strychnine alkaloid. And once again, plsease note - I do not recommend that this bait be used. Ask the folks where you buy it to remove the poison and sell you only the empty container, or you can throw away the poison (carefully) and wash out the container. Wear plastic gloves! Please be very careful with this stuff! Better yet - **just don't use it!** Instead, use a Hot Barrier.

Step One:

OBSERVATION AND LOCATION

◆ What plants are being affected by the gophers?

◆ Are there any signs of them being around? Mounds?

◆ Is the lawn being damaged by them? When you have spotted the telltale signs of gophers or moles (a small mound of soil or dead plants), locate an entrance to their tunnels[6]. Step back and try to see any patterns at work behind the placement of the tunnels. Often times this will be obvious. See if you can determine a source[7]. If you live in an environment where there are lots of gophers you should start thinking in terms of protecting the entire property, otherwise you'll concentrate on specific sites.

◆ Mark all active tunnels that you find with bright flags.

Step Two:

◆ Obtain the injector tool[8]

◆ Obtain or make the Hot Barrier (following)

◆ Obtain or make the Gopher Balls (following)

Step Three:

Inject the Hot Barrier into the tunnels: once you have located and marked off the tunnels then you can start to inject or place the following into the tunnels:

THE HOT HOT MIXTURE:

◆ 1 lb cayenne pepper or any hot chili etc. (the hotter the better)

◆ 1/4 lb garlic powder

Mix well and keep dry. Place the mixture into an active tunnel using the injector[9], then cover the entrance. Cayenne pepper [10] is a natural irritant and gophers or moles will not like it. Garlic, peppercorns, etc. will work well, too. A good device to use to inject the mixture into the gopher tunnels is the Gopher Getter (Midget and Jr.)*. The Gopher Getter Midget is a smaller unit which holds a teaspoon of product, while the Gopher Getter Jr., is a larger,

---

[6] There are many entrances to their tunnels. You can observe them as they pop up to look around. Usually a small mound tells you that they are there and that's the front door.

[7] They tend to live in a 'main' area and work their way around from there. Usually the main area is near a source of water.

[8] These can be purchased from any garden mail order catalog such as Peaceful Valley, Gardeners Supply, Harmony Farm Supply, Mellingers and lots of others.

[9] If you don't have an injector, you can use the pole end of a broom to make a hole into the tunnel. Then remove the pole slowly and pour the mixture into the tunnels. Cover the hole gently to avoid caving it in. A light pole is better then a heavy metal pole since you will need to feel your way into the tunnel. You will develop a 'feel' for when you have entered a tunnel.

[10] Or another 'Hot Barrier', see Hot Barriers in the Pest Control chapter.

heavy-duty commercial model which holds 24 ounces of materials. These models are available from many mail order catalogs. They are lightweight and easy to use. The baiter applies bait directly into the tunnel by pushing it into the gopher runway.

Hot, Hotter, Hottest!

This hot mixture, when placed into all of the tunnels, starting from the center and expanding out towards the edges of your property, will keep gophers and moles away as long as the mixture stays dry. It will work for 24 hours up to a week at a time depending on the weather and other conditions. See the Mo Hotta Mo Betta catalog for more hot chilies and other hot stuff you can use. Also see Hot Barriers in the Pest Control chapter. Don't water for 24 hours after treatment.

Hominy Grit 'Em

You can alternate Hot Barriers with hominy grits in the tunnels. The grits must also be kept dry to work. They can either be mixed in with the hot mixture above or used straight. Any instant grain will work.

Using Smell

Hair today, gone tomorrow! A good 'gasser' is decomposing hair. A strong odor is given off as the hair decomposes. Placing human hair in the gopher tunnels will effectively keep them out of passages and should be placed in new tunnels as they appear. Hair works best on hot days and hot nights: the heat causes the hair to smell up the tunnels! Get your hair from "the good ole boys" barber shops instead of hair from a salon since most men spray less 'stuff' on their hair[11]. If you can, use animal hairs such as dog or cat since this will repel the gophers even more effectively. Visit your local pet parlor!

Gopher Rocks: Straight from Europe, a product called Rodent Rocks are porous lava stones that have been soaked in an herbal formula containing garlic and onion. While not necessarily considered gassers, the rocks do smell up the tunnels. When buried, the rocks give off a very strong odor that effectively repels most rodents (including gophers, moles, and mice). Bury the small rocks (six inches deep and two to four feet apart) in a protective circle around your vegetable garden or problem area. Rodent Rocks remain effective for four to twelve months. This product is widely used in Germany. I recommend this in place of poison bait if you have animals. Using fish heads in gopher tunnels will irritate them into moving on!

Holy Smelly Cotton Balls Batman! A cotton ball soaked in either citronella oil, garlic oil, or crushed onions and placed into the gopher tunnels will disgust them enough to make them leave the area! Wear gloves to disguise your 'human' scent. Cover up the entrance and pat down the soil. Repeat at all the locations where gophers or mole 'mounds' are present. Return the next day and find the tunnels out of which the cotton balls have been tossed. That is an active tunnel. Throw the cotton balls back into the tunnels and cover the entrance with a large rock. This will last anywhere from a few hours to a few days. They will be back once the smell has disappeared.

---

[11]Nothing personal but many women spray their hair with many things which don't decompose too well. On the other hand this can prove to be a plus since some hair products stink and the fact that they don't decompose could mean that they will last longer under adverse conditions! So just use any hair you can find!

"Use anything natural that has a foul odor. Cotton balls soaked in ammonia or garlic will work well. Dirty socks any one?"

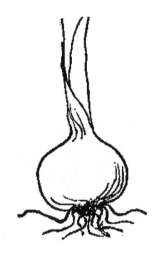

More Stinky Stuff: There are many products on the market which are used to keep deer away from your plants. These same products can be used on the gophers! Just dip the cotton balls into this stuff and throw them into the gopher tunnels. These products should last for a few days up to a few weeks! Try Deer Off. Just make the mixture and instead of spraying it to protect your plants from deer, etc, you place cotton balls to soak then place cotton balls inside tunnels!

Egg 'Em: A rotten eggs mixture can be made from (you guessed it) eggs! This is easy enough to do. Buy a carton of eggs. Take one egg out at a time and carefully place into a gopher tunnel, then cover up the entrance. The egg will rot (sooner then later) and when the gopher comes to investigate and breaks it open...oh boy!

Step 4:

USING BARRIERS

The use of underground barriers is recommended in controlling gophers/moles around your home. Use aviary wire that is woven tightly enough to keep the gophers from going through and strong enough to resist their chewing (no larger then 1/4" mesh). Place the wire underground, around an area where you plan to plant a garden or flower beds. You can also use root guard baskets. These are pre-made gopher baskets made from wire mesh. There are many sizes available or you can make your own. Try Harmony Farm Supplies, Gardeners Supply, Peaceful Valley, or Arbico to name a few. See the resources section.

Another homemade barrier consists of bottles placed into the ground. Tall wine bottles are great for this! Just dig a trench and place bottles standing up as a barrier around your garden, etc. The gopher or mole sees its reflection and runs away - plus, they can't dig through the bottles. I have also used everything from wood to rocks to broken glass as barriers.

You can also use plants that gophers naturally abhor. Garlic and onion plants are avoided by gophers/moles and make excellent barriers. Other plants gophers hate are herbs such as tansy, peppermint, spearmint and rosemary.

Step 5:

ENCOURAGING PREDATORS

The use of animal predators is the one of best ways to naturally control gophers. If you have animals, the use of poisons is not recommended. Cats and dogs will chase gophers and sometimes eat them. Other predators that will eat your gophers for dinner are king snakes, gopher snakes, hawks, coyotes, and owls. Try buying a gopher snake or king snake from your local pet store and releasing it down the gopher hole! A king snake is also a natural enemy of the rattlesnake. Or try building a nesting home for owls and other birds.

Step 6:

USING PLANTS

Gopher Purge - uphorbia lathyrus. Each pod contains three seeds. It will repel gophers and moles and other tunneling and burrowing animals.

> *"There are many products on the market which are used to keep deer away from your plants. These same products can be used on the gophers!"*

The roots are extremely caustic to them as well as to humans—so be careful not to get the white milky stuff on your hands or face; if you do, wash it off right away. Gopher purge is a parental and gives plenty of seeds for replanting. Water it well at the start. The whole plant can also be dried and placed into the gopher tunnels; there are many other uses for it, too. For maximum protection it is important to plant gopher purge as a hedge that surrounds the property. Plant gopher purge (a complete seed pod should be planted) every six feet the first year, decreasing it to every three feet the second year, and then down to every two feet if the gophers are still bad.

Step 7:

USING SOUND WAVES

There are many ways to use sound to tell the gophers that you want them to leave. One way is to place a speaker on top of the ground on the area where you don't want them and play something through it really loudly! Another thing that works is to buy or make your own clanker! What is a clanker? It's a wind driven device which can be made to look like anything. As the wind turns it, it makes a sound which is transmitted straight down through a pipe into the gopher tunnel. Eventually the gophers will get used to it.

In my next book I'll tell you how to make a small radio into a big gopher remover! Okay, okay - just buy yourself a small battery-powered radio and place it into a gopher hole. Allow it to play until the batteries run out and replace them as needed.

Gopher It Battery Gopher Getters are placed into the ground to protect vegetable gardens and small lawns. They give off ultrahigh sounds to drive gophers and moles away. They cover a 1,000-square foot area. This product offers a safe method of repelling rodents (gophers, moles, ground squirrels, and pocket mice) with no damage or danger to the ecological balance. Go'pher It does not kill but repels. They are available from many catalogs. Try Gardeners Supply, Sharper Image, etc.

Uphorbia Lathyrus (Gopher Purge) This is the safest method of ridding yourself of gophers. Give it time; it works!

Step 8:

USING TRAPS

One of the most effective ways to get rid of gophers is to trap them (be aware - many of these traps actually kill them). One of the best traps around is called **The Black Hole** (a black circular trap with a hole at the end in which you place an apple attached to a cord). Ask for it by name. Here are a few of the other traps available on the market and a brief description of each:

Gopher Traps

◆ **Cinch Gopher Trap** has a metal plate with a cinch lasso which catches the gophers.

◆ **Guardian Gopher Trap** is a box type trap with a spring trap that kills them.

◆ **Macabee Gopher Trap** is a spring-type, all wire trap (see left).

◆ **Cinch Mole Trap** is the same as the cinch trap for gophers but smaller.

◆ **Out of Sight Mole Trap** is a scissors jaw and spring type of trap that has been in use for a long time.

Step 9:

Using Poisons

Please use poisons only as a last resort. See the Pest Control Chart for other alternatives. **Poisons** are not always the best answer to this problem. Try the eight steps above before you try this step. Even then, avoid using any standard poisons since they will kill other animals as well.

Sulfur: Gassers that are often used to evict gophers from your property are made of sulfur and potassium nitrite. They are lit and placed in the tunnel, producing large amounts of smoke which drive the gophers/moles away. They must be used regularly to be effective. Avoid using this method unless you know how to use the gasser.

Great care must be exercised because the sulfur smoke is very dangerous to inhale and will cause serious illness. It is available at most nurseries. Avoid using sulfur if you have children or dogs.

I have used cigars, lit and placed into tunnels. Cigarettes work well, too. Avoid using any other kind of poisons!

D-3 Vitamin D: D-3 is a form of Vitamin D that kills gophers, rats, ground squirrels, and mice. Rodents have trouble regulating calcium in their blood stream and vitamin D disrupts their systems, leading eventually to death. Using vitamin D is safe for humans, household pets, other animals, and birds. Add it to your injector and place it into tunnels—but only after other attempts have failed. It is available in pellets or granules from most mail-order catalogs like Peaceful Valley or Harmony Farm Supply. It is sold under the trade name of **Rampage**. Please use this only as a last resort!

Using Tobacco: Gophers do not like tobacco smoke any more then we do. When inhaled, it can cause serious damage to both humans and gophers! When eaten by gophers, tobacco leaves can kill them. Tobacco is easy to handle as it comes in cigarettes, cigars, and pipe bags. Try to buy an organic variety or one that uses the least amount of chemicals. In most cases you can just buy a pack of cigarettes (non-filtered) and place one cigarette into each tunnel. The gophers will either eat these and die or go away and leave you alone. You can also light the cigarettes and then place them into tunnels.

Castor Oil Anyone?: Castor oil is a very useful oil to use in controlling many different types of pests in your garden. It is especially useful in controlling gophers, moles, and voles in your lawn or garden. Take a cup of castor oil to which you have added one gallon of hot water. Pour this mixture into the gopher hole. Make a second batch and pour it around the effected area. Alternatively, you can pour the castor oil into a sprayer that is attached to the garden hose. Get a sprayer that allows you to control the oil to water ratio. Adjust to the highest ratio possible. Miracle Grow makes a great spray container that attaches to the hose. Put the castor oil into this container and spray the entire area that has the gophers. The best time to apply is in the early part of a good hot day. Apply again in a week and then apply every month until the problem disappears.

"How can you tell whether you have a gopher or a mole? Gopher mounds have dirt pushed off to one side of the mound while a mole mound will be volcano shaped with the center at the entrance to the tunnel (usually plugged up)."

When there is balance there is co-existence.

4

Organic Fertilizers

Growing pest free vegetables and fruits is in part dependent on not using chemical fertilizers and, instead, relying upon the various natural resources available to us. It is not within the scope of this book to debate the difference(s) between organics and inorganics; nor is it my intent here to show why you should not use chemical fertilizers or if a plant can tell the difference between organic nitrogen sources and chemical sources[1]. I am not here to convince you that you should go organic. This is something you must decide on your own. I am here to pass along to you my knowledge of Organics.

In this chapter, I will cover some basic methods of making your own organic fertilizers. Please note that while there are more and more organic fertilizers available on the market today, not all are 100% organic. Read the labels very carefully. The physical structure of what we feed the earth can be modified. For example, a rock can be crushed into powder, or garlic made into a liquid. The point is, all these things are still subject to the laws of nature. They are biodegradable and non-toxic.

High nitrogen-based fertilizers are to the earth as sugar to the human body. The rush is great going in, but coming down—watch out! Each time you use high nitrogen, it causes greater stress and greater stress to the plant life. Stressed plants attract pests and diseases. High nitrogen is like steroids for plants!

The Invisible Gardener says: "You can avoid many problems associated with high nitrogen use simply by understanding the organic system of providing nitrogen as the plant needs it."

Organic Elixirs

Here are some formulas for organic fertilizers. I have included them to help you develop a feel for blending your own organic materials. Remember that the numbers only signify what is available immediately, not what will become available later on, as nutrients are constantly released into the soil by the action of the microorganisms in the soil. Also, you may adapt these formulas to suit your needs and experience[2]. Any product marked with an * is available at NITRON INDUSTRIES, ARBICO, GARDENERS SUPPLY, ACADIE SEAWEED, C. P. ORGANICS, ALBRIGHT SEEDS. More information is available in the resource directory. (Not listed? Give us a call for information on becoming an IGA commercial member or visit our website at: http://www.invisiblegardener.com/joiniga.htm)

---

1 Which it can.

2 If you don't need a lot it's best to make small batches.

For Fruit Trees:

The following recipe makes approximately 30 pounds[3].

2 lbs fish meal

4 lbs ord ashes

1/2 lb trace minerals *

2 lbs rock dust *

2 lbs seaweed meal (like Acadie) or kelp meal *

2 lbs coffee grounds

10 lbs compost *

5 lbs composted animal manure

Blend all ingredients together thoroughly. Make only enough for immediate use. Keep dry. Spread a thin layer under each tree, starting one foot out from trunk and extending to twenty feet past the drip line. Water well into soil. Use approximately one pound per year of growth. A ten-year-old citrus tree should get ten pounds a year: five in the spring and five in the fall. Apply lots of compost or horse manure in early spring[4]. Always add a soil acidifier such as pine needles or leaves to the compost or to the horse manure in order to lower the ph level. Most fruit trees prefer 6.5-6.8 ph. Never use freshly cut grass clippings as a mulch[5]. Spray leaves with a liquid seaweed solution (like SuperSeaweed or Nitron A-35 or Agri-Gro or Roots Plus or Acadie Seawed) or with a seaweed mixture of your own making[6], once a month. Adding small amounts of a natural soap will control pests. See soap page.

For Roses

The following recipe makes 30 pounds. To make less, cut amounts porportionately.

2 lbs fish meal *

4 lbs organic alfalfa meal *

1 lb organic cottonseed meal* (if soil is alkaline)

1 lb wood ashes (if soil is very acid)

2 lbs rock dust*

4 lbs kelp meal or seaweed meal (like Acadie)*

1 oz garlic powder

10 lbs compost *

5 lbs composted animal manure

Blend all ingredients thoroughly. Use once a month during main flowering season. Use one cup per plant, per month. Water well into the soil, without wetting leaves. The compost should have an acidifier added to it, such as aged wood, pine needles, etc., to bring the ph down. Roses grow best at a level of 6.5-7.2 ph. Roses love lots of rich compost and mulch. In early spring, add enough compost to fill well around roses. Water in well before mulching with leaves, bark etc. Roses grow best with a drip system. For best results, spray weekly or monthly as needed with a liquid seaweed or liquid

The Invisible Gardener says: "High nitrogen fertilizers contribute to insect & disease attacks."

enzyme (SuperSeaweed, Nitron A-35 or Agri-Gro, Acadie) and natural soap mixture (such as Dr. Bronners soaps). See chapter on making your own pest controls for additional rose sprays. Check out C.P. Organics Seabird Guano as an excellent organic fertilizer you can add to roses. Use one cup per month to increase blooms and growth. Also check out Nitron's Nature Meal for roses. This is also an excllent product you can either add directly to your roses at one per cup per month or add to the mixture above. Try also using Albrights Biosol.

For Azaleas

The following recipe makes about 25 pounds.

4 lbs organic cottonseed meal*

2 lbs rock dust*

4 lbs organic alfalfa meal*

1 lb kelp meal or seaweed powder (like a Acadie seaweed)*

4 lbs organic coffee grinds (fresh coffee can be used)

10 lbs compost (mixed with equal amounts of aged wood)

Blend all ingredients together and use as needed. Azaleas love an acid-low ph level of 5.5-6.0 for best results. Always add an acid-aged wood to your compost when feeding azaleas. Avoid using peat moss since the supplies are getting low. We should find alternatives to peat moss and allow the peat bogs around the world to return to their intended state. Try using coconut husks as a substitute. C. P. Organics Seabird Guano is also good for this.

For the Lawn

Makes about 45 pounds.

10 lbs fish or kelp meal (like Acadie or Maxi Crop)*

10 lbs organic alfalfa meal*

5 lbs composted animal manure*

10 lbs rock dust*

1 lb wood ashes (if available)

10 lbs compost*

llama pellets finely ground.

Blend all ingredients together and apply four times a year, or as needed. Sprinkle thin layer over lawn and water well. Lawns love compost; top-dressing lawns once per year with a good rich compost will keep the soil alive

---

[3] The formula here is just a guide line for you to follow. Use what you can get ahold of first. Don't stress yourself out looking for something on the list. If you can't find it, don't worry! Use something like it instead or ask your local nursery, or call up my web site for free! (See info in the back of this book.)

[4] By lots of compost I mean approximately 50 lbs of compost per mature tree. 100 lbs would not be too much for a 10 year old tree.

[5] Grass clippings, though full of nitrogen, don't make the nitrogen available to the soil/plants until it has broken down. During the breaking-down process, the clippings will actually take away food from the soil and plants. Grass clippings are best used as a light mulch when there is plenty of good rich compost already available.

[6] See the chapter on making your own foliar spray.

What's in 'Organic Food'?

Everyone wants organic foods. Sales of $173 million in 1980 have exploded to $3.5 billion a year now. What are those growers using for fertilizer? Not until Secretary of Agriculture Dan Glickman released the nation's first regulations for organic food in 1997 have consumers of organic foods been able to know exactly what they're buying. The new regulations are a chaotic patchwork of the labeling laws that states are currently using or want implemented. The confusion, however, is not entirely over, for rather than accepting the carefully deliberated recommendations of the National Standards Board, Glickman declined to definitively prohibit three practices being used in producing organically grown foods.

For the record, the Invisible Gardener believes that the secretary should ban each of these practices:

FERTILIZING WITH SEWAGE SLUDGE.

When cities were barred in the 1970's from discharging the runoff from their sewer systems, some people began selling the materials to farmers as fertilizer. The use of sewer sludge in any kind of agriculture is questionable, given the toxic waste it often contains, including cadmium and lead as well as other heavy metals and toxins.

IRRADIATION.

While irradiation appears

(Cont.)

to be a safe way of preventing hazardous bacteria from infecting beef and poultry, it has not been properly tested and may yet prove to be harmful to humans and the environment. Organic foods are just as vulnerable to these bacteria. The National Organic Standards Board, however, sensibly argues that most Americans would not expect food advertised as organic to be irradiated. Irradiation is just a clever way of getting the public to allow their food to be exposed to anything from toxic wastes to nuclear wastes!

GENETIC ENGINEERING.

This is a growing practice wherein, for example, a bacterium gene is injected into tomatoes, turning them into 24-hour-a-day anti-pest factories. This yields a large volume of unspoiled tomatoes in the short term. But in the long term insects develop resistance to the super-tomatoes, leaving farmers scrambling to find ever newer strains of protective bacteria. This, too, has not been properly tested (actually the US Goverment decided that it was safe without testing it and, furthur, decided that genetically engineered produce does not have to be labeled as such). This is a violation of the public's right to know.

MAIL COMMENTS TO:
National Organic Standards Program
USDA/AMS/TNM
2510 South Building
PO Box 96456
Washington DC 20090-6456

and allow for deeper root systems and a pest-free lawn. Compost should be raked into lawn and then watered well. The best time to apply compost to the lawn is in early spring (or as soon as the last frost is over). For best results, spray your lawn monthly with a liquid seaweed such as Superseaweed or Nitron's A-35 or Agri-Gro's Turf Formula, or with Gardeners Supply's Roots Plus for Lawns or Acadie Seaweed. There are many natural fertilizers available on the market today. These are just a few examples selected from IGA commercial members' catalogs, because I know these products and feel confident in recommending them to you. But I strongly suggest that you be careful of the products on the market today and that you always ask questions and read the ingredients. Is it 100% Organic? What else is in it?

For African Violets
Makes about 8 pounds.

1 lb organic cottonseed meal *

1/4 lb rock dust *

5 lbs compost

? lb organic alfalfa meal *

1 lb kelp meal or seaweed powder (such as Acadie) *

Blend all ingredients together and apply using a tablespoonful per small container of plants, or add a tablespoon of the mixture to a cup of water and sprinkle it on the violets. They also love Nitron A-35, SuperSeaweed, Agri-Gro, and Acadie seaweed.

For Container Plants
Makes about 13 pounds.

10 lbs compost*

1 lb rock dust*

1 lb organic alfalfa meal *

1 lb kelp meal or seaweed powder *

Blend ingredients together thoroughly and apply mixture as needed to container plants. Usually one cup per five gallon container will do, but the exact amount will depend on the size and the age of the plant. Spray regularly with either SuperSeaweed, Nitron A-35, Agri-Gro or Acadie seaweed. Use C.P. Organic Seabird Guano for containers.

For the Vegetable Garden
Makes about 40 pounds.

10 lbs compost *

5 lbs rock dust *

10 lbs kelp meal or 10 lbs seaweed meal *

10 lbs composted animal manure*

5 lbs organic alfalfa meal *

Blend all ingredients together. Feed plants weekly or monthly as needed during growing season. Vegetable gardens love good rich compost. Apply the compost as both a top dressing and a mulch.

Compost and Mulch
For Sick Plants

10 lbs compost *

2 lbs rock dust *

2 lbs organic alfalfa meal *

2 lbs kelp meal or seaweed powder (like Acadie)*

Avoid high nitrogen chemical fertilizers[7]. Spray with a liquid seaweed concentrate such as Superseaweed, Nitron A-35, Agri-Gro's Premier Plant Food, Gardeners Supply Roots Plus or Acadie Liquid Seaweed. Apply it as often as needed. You can also use C. P. Organic Seabird Guano or Albrights Biosol for good results in recovering your plants. Apply lots of compost!

Sources
Composting Materials or Readymade composts

Local restaurants, especially the natural food ones (their vegetables, etc. are more likely to be organic). Use their leavings in your compost.

Seafood restaurants or fishing docks to obtain seafood to be dried for your compost use.

Clam shells and other beached bits. When dried and crushed, clam and other shells make an excellent addition to a compost pile.

Coffee and tea. Look for any sources of coffee such as either coffee hulls and or coffee grinds; similarly, locate sources of tea, either tea bags or leaf.

Lumber yards may also be a good source of material, provided that you use only the detritus of untreated wood products.

Chicken farms. Locate chicken farms and ask for chicken feathers, egg shells, or droppings. (Ask them what they are spraying, if anything.)

Horse or cattle farms. Locate horse or cattle farms and, after asking if they spray their animals or the manure, use the manure.

Organic mushroom farm. Make sure it's one that uses steam instead of chemicals.

Old alfalfa bales.

Grass clippings make an excellent source of nitrogen for your compost pile, just make sure it's organic (unsprayed).

---

[7] Chemical fertilizers, especially those with a high nitrogen base such as Urea, do considerable damage to the soil, eventually killing off all beneficial bacteria which then leads, you guessed it, to sick plants!

[8] Using bacteria, toxins can be removed from the soil. This is the future of composting plants and sewer plants. Until then ask questions concerning your local sewer sludge. Sewer sludge and compost don't go well together. Unless the sewer sludge is processed using bacteria, it will contain heavy metals and other toxins that will end up in your environment, in your food, and inside of you!

DON'T WANT TO MAKE YOUR OWN COMPOST?

Look around your city. Many cities are making compost these days. Ask what it is made from. Avoid ones with sewer sludge unless it is clean of heavy metals, etc. See if you can spot any places that sell compost. Ask other gardeners for a source. Go out into the countryside and look around there. If you find someone making compost, always ask them how they are making it, if they are exposing it to any chemicals, how long they have been making it, etc. These are important questions. (And never assume that anything is as someone says it is.)

Nowadays there are more and more people making and selling compost, so it should not be hard to find.

Nitron Company, Arbico Company, Peaceful Valley Farm Supplies, Gardeners Supply all sell compost thru the mail.

Sewer sludge can be processed correctly with an end result of good, clean, safe compost[8]. However, studies around the world have shown that such a process is not the one used by most of the world's sewer treatment plants. Especially in the USA. Cont.

COMPOST? CONT.

This is an area that we must look into very closely if we are to solve this problem. Thru the use of bacteria, we can make sewer sludge a safe organic alternative to fertilizers. The laws are changing rapidly every year. We must keep close watch over this or we will have a problem that will be very difficult to resolve. Do not believe it when manufacturers say the toxin levels are below EPA standards. Insist that they lower the standards to zero levels. Remember every little bit of toxin adds up to a lot. Check with the source before you buy it!

"Nothing beats organic

alfalfa tea"

...the Invisible Gardener

For more information on compost see the compost chapter.

Some examples of organic fertilizers (only IGA commercial members are listed here[9]).

ORGANIC SUPPLIES AVAILABLE FROM **ARBICO:**

Ion 2000. This is a totally organic fertilizer derived from mined minerals and plant extracts. It comes as a soluble powder. There are two formulas: 16-15-16 and 10-45-10.

Gro-Up Fertilizers. Made from crawfish, crab shells and cricket castings. The only organic fertilizer which has an organic nematocide!

Plant and Garden Food 4-3-1. A specially blended organic fertilizer for tomatoes and vegetables.

Tomato and Vegetable Food 4-4-4. Great for the vegetables in your garden.

Blooming Plant Food 4-8-4. Helps your vegetables produce flowers.

Organic alfalfa meal[10]. Organic alfalfa meal has been a long favorite of mine. This is one of the best sources for microorganisms plus trace minerals, NPK, and more. It's excellent as a tea for foliar spraying. Can be used for roses, vegetables, fruit trees, lawns, etc.

Bat Guano. This product has both a high nitrogen and a high phosphorous formula. Makes an excellent manure tea for foliar feedings!

Calcium 25. Used as a foliar feeder for calcium. It contains 25-30% calcium from natural salt rinds and also has trace minerals. Combine this with Nitron A-35 for excellent results.

Compost Power. This is Arbico's best compost. If you can't make your own try this.

Organic Cottonseed Meal 7-2-2, For acid loving plants.

Earthworm Castings. These little guys really produce a great natural fertilizer. You can also get the earthworms.

Fish Meal 10-2-2. This is a great natural fertilizer for all plants. Great for vegetables, lawns, roses, trees.

Gypsum Power. An excellent source of calcium and sulfur. Helps to regulate PH levels, adds calcium and sulfur to soil and leaches out excess magnesium and sodium.

Soft Rock Phosphate. 27% phosphate, 16% phosphorus, plus 18 trace minerals.

Sul-Po-Mag. 22% sulfur, 22% potash and 11% magnesium.

Kelp Extracts. They carry a wide variety of kelp extracts such as Algrow brand, Maxicrop brand, Shur Crop brand, and Nitron's kelp meal.

Liquid Kelp: They carry Sea Grow (my favorite), lime soil conditioner, liquid iron complex, natures humic acid as well as Nitron A-35. See Nitron for more info.

Trace Elements: A special liquid formula containing iron, manganese and zinc. A great foliar feeder. Use with Nitron A-35 or SuperSeaweed.

Biotron Soil Inoculants: Contains over 25 strains of beneficial bacteria, fungi, enzymes, algae and yeasts. Enhances nitrogen fixation.

Bio-Dynamic Spray: A BD compost starter that's designed as a foliar feeder for your plants.

Organic supplies available from Nitron Industries:

Nitron A-35: "We knew Formula A-35 had transformed our Kansas farm in 1977; others had to tell us there was nothing else in the world that worked so well! The first hint was the awestruck expression on the face of garden writer Pat Branin when he came to our farm to see for himself. He went home to test A-35 in his own soil, then he wrote an article for the San Diego Union that brought the world to our door. We had been sharing A-35 with friends and neighbors, sending them off with a fruit jar or bucket full; but when phones started ringing and we were talking to growers from around the nation and the world who desperately wanted help in reclaiming poor soil, there weren't enough fruit jars!" Frank and Gay Finger[11].

Nature Meal 9-3-6 for Lawns: A natural fertilizer that helps build strong, healthy lawns.

Nature Meal 8-2-2 all purpose: Feed everything you have growing!

Nature Meal for Bulbs: Contains a greensand mineral base.

Nature Meal for Vegetables 4-8-4: Contains over 96 trace minerals, plus a protein and carbohydrate base for sustained plant growth.

Nature Meal for Tomatoes 5-6-3: Studies by researchers from major universities across the country are merged in this product, which provides exactly the right ratio of minerals, proteins, carbohydrates and vitamins required to grow picture-perfect tomatoes.

Nature Meal for Roses & Flowers 4-6-2: Andy says: "Nothing works better for roses then Nature Meal. Try it and you will agree!"

Kelp Meal: One of the best on the market. High in the right stuff.

Fish Meal: A great slow release fertilizer, takes 30-60 days to be converted into nutrients for your plants. Great for compost too!

Fish Emulsion Fertilizer[12]: This is probably the only urea-free fish emulsion on the market. They have developed their own method of pro-

The Invisible Gardener Says: "Avoid Buying Compost made with sewer sludge since it may contain heavy metals which can pollute us and our environment."

---

[9] The Invisible Gardeners of America publishes this book for use by its members and the general public. Commercial members include any businesses providing a service or product that promotes environmental awareness and environmental actions such as the five R's: Reduce, Re-Use, Re-Cycle, Re-Think and Re-Plant. Once you become a commercial member (not all are accepted) we will test your products in our organic gardens and then I will include it in our next edition. My members trust my knowledge in this field and have come to look forward to the latest edition of this book to see what's new. I will not write about something that I have not used myself. Visit our web site at http://www.InvisibleGardener.com/joiniga.htm for more info on joining or see the back of this book.

[10] Nothing beats Organic Alfalfa tea.

[11] I myself have been using Nitron A-35 since 1978 (I answered one of their first ads) and I can tell you that there really is nothing like it. Nitron A-35 has become a great part of my garden. I really enjoy watching the plants, lawns and earth respond to the A-35.

[12] They make one of the few urea-free fish emulsions on the market. This fish emulsion, because it is fermented, is a very powerful fertilizer/pest controller. The oils in this product alone control many insects.

cessing which makes it almost odor free. Great foliar feeder. Feeds lawns, roses, flowers, vegetables.

Sea Grow 1-1-1: One of the best liquid seaweeds on the market.

Sea Grow Plus 2-1-2: Blend of plant foods from the ocean combined with micronutrients - nitrogen, phosphoric acid and potash - of ocean fish emulsion, and the micronutrients, trace minerals and amino acids present in liquid kelp extract. Best foliar spray around. Use with Nitron A-35 for even better results.

Compost made from poultry manure: This is a great compost product if you can't make your own.

Nature's Own Humic Acid: A natural product mined from an ancient deposit of leonardite ore that has decayed over millions of years. Provides bacterial activity to soil. Excellent rx for ailing house plants.

Earthworm Castings: Nothing better then what the earthworm produces. Rich in trace minerals and more.

Feather Meal 13-0-0: A long lasting high nitrogen natural fertilizer, feather meal provides plants with high energy longer.

Organic Cottonseed Meal 7-2-2: A great soil acidifier. Acid loving plants love this stuff! Also great for the compost pile!

Organic Nitrogen: A blend of three natural nitrogen sources, count them: Feather meal, blood meal and fish meal. For all your vegetable, and flowers, lawn.

Blood Meal 12-0-0: A slow release natural nitrogen, blood meal stimulates bacterial growth with just a sprinkle! Also works well for keeping rabbits and deer away. Add it to your compost pile to speed up the process and also to provide much needed nitrogen.

Bone Meal Plus: 100% organic Bonemeal Plus is made by the old fashion method of steaming. Another interesting thing is that these folks process this stuff with meat still on the bones for added protein. Contains 25% calcium, 12% phosphorous and 5% nitrogen. Great for your bulbs, fruit trees, roses and, of course, to add to your compost pile!

Organic Alfalfa Meal and Alfalfa Pellets: Not enough can be said about this stuff. Provides vitamin A, plus thiamine, riboflavin, pantothenic acid, niacin, pyridoxine, choline, proline, bentaine, folic acid, N-P-K-CA, Mg, and many other valuable minerals. Also included are sugars, starches, proteins, fiber, co-enzymes and 16 amino-acids. How could your soil not respond to such a nutritional meal? (Heck, I would eat this stuff.)

Greensand: An old time favorite of organic gardeners, greensand is a natural marine deposit containing potash, silica, iron oxide, magnesia, lime, plus over 30 trace minerals. It's a very slow time-release fertilizer.

Soft Rock Phosphate: 27% calcium, 16% phosphorus plus 18 trace minerals. Helps rebuild soil structure.

Sul-Po-Mag: 22% suplhur, 22% potash and 11% magnesium. 100% water soluble, non-acid forming and does not increase soil acidity.

Bat Guano 3-8-1: The folks at Nitron went out of their way to locate

---

"I myself have been using Nitron A-35 since 1978 (I answered one of their first ads) and I can tell you that there really is nothing like it. Nitron A-35 has become a great part of my garden. I really enjoy watching the plants, lawns and earth respond to the A-35."

The Invisible Gardener

a good source of this stuff just for you! Straight from Montego Bay Jamaica! Ummm Good!

Rock Dust: This stuff is so good that I wrote a whole chapter on rock dust and its many uses. See the Rock Dust chapter. It comes from an incredibly rich glacial sea sediment. 5% sulfur, 5% calcium and over 30 trace minerals.

Water Soluble Fish Concentrate 12-0-0: Now you can make your own foliar spray. Great for lawns, vegetables, flowers.

Water Soluble Blood Meal 14-0-0: If you have ever tried to dissolve blood meal then you know why this is such a great product! indoor magic for all your house plants, contains A-35 plus natural proteins, plant growth regulators and minerals.

ORGANIC SUPPLIES AVAILABLE FROM LIVING WATERS

AGRI-GRO: Agri-Gro is another of those great products that we humans invent! Agri-Gro is an enzyme product that helps restore life to

the soil. It combinds aerobic and anerobic bacteria enhancers that produce nitrogen from the air and enzyme action in the soil. I love this stuff! Agri-Gro is not an NPK fertilizer but it works to enhance the relationships between soil and plant. See resources and index for more info on this product.

Turf Formula for Your Lawn: Helps your lawn to grow healthier by making better use of the bacteria in the soil.

Premier Plant Care Formula: This formula is great for your indoor plants as well as for container plants.

ORGANIC SUPPLIES AVAILABLE FROM ALBRIGHT SEED COMAPNY

Biosol Plus 7-3-3: A natural all purpose fertilizer made from soybean meal, organic cottonseed meal and sulfate of potash magnesia. Contains no chicken or animal manures yet is an effective biostimulant and natural chelating agent.

ORGANIC SUPPLIES AVAILABLE FROM GARDENERS SUPPLY

Roots Plus 5-3-4: Roots Plus is an organic soil builder and organic fertilizer in one product. A good example of what modern high tech can do.

Roots Plus 12-2-3 for Grass: Great for keeping your lawns nice and green longer.

Roots Plus 3-5-6 for Vegetables: Roots Plus for Tomatoes 5-6-5

Earth Worm Egg Castings: Grow your own fertilizer factory.

Gardeners' Supply's Own Organic Fertilizer 5-5-5: Contains naturally occurring minerals, peanut meal and whey meal.

Bat guano makes an excellent manure tea for foliar feedings!

KELZYME ANALYSIS:

ELEMENTS % OR PPM

aluminum 5,089.00

arsenic .... <75 .42

barium 90.37

boron .... <1.80

cadmium .... <1.5

calcium 32%

chromium 3.13

cobalt................... <2.6

copper T2.82

iron1 .... 1.25%

lead .... <15.

lithium 31.15

magnesium 4,638

manganese 3 16.72

mercury .... <1.

molybdenum........ <1.7

nickel 33.37

phosphorus........ 580.8

potassium ... 1,3 10.85

selenium .... <24.6

silicon 1,399.68

silver .... <0.80

sodium .... I, 040.71

strontium 278.94

sullur 2,256.95

tin .... <14.O

titanium 200. 26

vanadium.......... 41 .65

zinc 33.32

Gardeners Supply own Organic Lawn Fertilizer 7-3-4

Black Rock Phosphate 0-4-0: A slow release phosphorus supply. Neutralizes acidic conditions in soil.

Desert Bat Guano 8-4-1: A rich source of nitrogen and phosphorus.

Fossilized Seabird Guano 1-10-0: Great source of calcium, phosphorus, and trace minerals.

Greensand

Marine potash, silics, iron oxide, magnesia, lime, phosphoric acid and 22 trace minerals.

Energy Buttons 3-4-3: Composted manure without the hassles.

ORGANIC SUPPLIES AVAILABLE FROM ACADIE

Seaweed 2-2-17: Acadie liquid and powdered seaweed is one of the cleanest and purist around. With over 2% nitrogen and 50-60 trace minerals, this is one product you want to use. 2% phosphoric acid, 17% soluble potash, 55% organic matter and nothing else.

ORGANIC SUPPLIES AVAILABLE FROM C.P. ORGANICS

Seabird Guano 12-12-2.5-6-1.5: Total nitrogen—12.00%; 6.70% ammoniacal nitrogen, 0.10% nitrate nitrogen, 3.95% water soluble organic nitrogen, 1.25% water insoluble organic nitrogen. Available phospheric acid 12.00%, soluble potash 2.50%, calcium 6.00%, sulfer 1.50%. All nutrients are derived from seabird guano. Peruvian Sea Bird Guano is produced in pellet form, for the purpose of supplying a 100% organic fertilizer product, high in nitrogen and phosphorus, and containing a moderate amount of potassium, calcium and sulfur, with ease of application in mind.

ORGANIC SUPPLIES AVAILABLE FROM ENVIRONMENTAL HEALTH SCIENCE, INC

Kelzyme(tm): Kelzyme is a 100% natural, non-burning plant food which comes from ancient fossilized kelp beds. Kelzyme contains calcium and over 25 trace minerals, as well as cytokinin, a plant growth regulator. Using Kelzyme will promote synergism, help stimulate natural nitrogen-fixing bacteria, and enhance enzymatic activity in the soil. Kelzyme is easily assimilated by plants and is a repressor of harmful bacteria and fungus in the soil. Kelzyme helps to produce healthy green foliage and contains no unpleasant odors.

Another source of organic fertilizers is Peaceful Valley Farm Supply. See resources for ordering address. Tell them The Invisible Gardener sent ya!

5

Organic Pest Controls

"There is a story of how the miners would take a bird in a cage down with them into the mines. If the bird died then everyone would leave since something was wrong. The same is currently happening to us. The disadvantages of chemicals are well documented with the result being the depletion of top soil[1] and the extinction of many species of life; to the mutation of pests into greater pests, and the weakening of all living beings (environmental toxins cause eventual death and extinction). The depletion of the ozone layer is but one example of this. This is our BIRD, the earth is our cage. We must heed the warnings!". Instead what we are doing is to replace the bird! This is not the answer.

The increasing use of synthetic chemicals in our daily lives is causing an increasing imbalance in Nature. When nature is out of balance it causes stress which in turn causes disorder and chaos. It is during this period that diseases and pests will strike. The plant and insect kingdoms are but a mirror into our world. We see the results of our actions within the insect and plant worlds. How long can we ignore the chaos? It is for these reasons that natural non-chemical life-styles are very important in developing a sustainable life-style. This has created a demand for the knowledge of how it is done.

The science of Organics is the understanding of the delicate balancing act which Mother Nature does every day. It is understanding this balance that makes Organics work. Organics has been used in pest control and farming since before written times. Chemicals are synthetic, temporary and unstable. Organics are natural, permanent and stable. Which is the Fad? Although organic sprays and dusts continue to play a crucial role, biological controls are of equal importance. By attaining a thorough understanding of the relationships between plants, soil, animals, insects and humankind, we can begin to understand the deeper relationship we all have with each other. This is the way the ancient farmers did it: by relying on natures' methods of maintaining balance.

Some Fundamental Rules of Organic Pest Control

1. **High energy soil provides greater balance that reduces stress and reduces the pest activity.** All Pest activity is linked to high stress levels. The greater the stress the less nutrition is available. Less nutrition increases stress! To put it in an easier way to understand; Energy can be of either a low or high energy.

When it comes to plants, they want the type of energy that can be converted into sugars that they need for the various plant functions such as assimilation and flowering. The higher energy sources provide to the plants a quick pick me up, Higher energy food sources are easier to be absorbed then lower energy sources which require more energy to assimilate leaving the

---

1 What's left is nothing but sterile and lifeless soil.

Rule #1

The Higher the

Energy,

The Greater the

Balance;

The Lower the Stress,

The Less the Pest.

"The frog does not

drink up the pond in

which he lives[2]"

- Buddhist proverb

plant in a minus situation which in turn causes stress etc. But if the plants are supplied with an abundance of food, a greater balance is achieved between the plant and its surrounding environment. It is through this balance that the stress of the plant is reduced. Reduced stress always equals reduced pests/disease.

2. Everything is linked together.

3. Providing a bacterial and mineral balanced soil insures correct nutritional availability to plants. Well-nourished plants develop stronger immunities to insect and disease attacks. Pests attack plants that are stressed out. See rule 1.

4. Always Strive to Achieve Ecological Balance in your Environment. Avoid doing any thing that causes imbalance. All Chemicals cause imbalance, upset the environment and increase stress. See rule 1.

5. Use only Natural Fertilizers and Natural Pest Control Methods. The Rule of Stress applies to all living things. Chemical fertilizers and chemical pesticides cause imbalance. Avoid high nitrogen fertilizers. Do not use Urea based products! Urea kills the beneficials, and stays in the soil for years. To have healthy soil you must have a diversity of bacteria and minerals, which is not possible when using chemicals. Chemicals kill earthworms, destroy bacteria, lock up minerals in the soil, and cause illness through weakening the biological systems, opening them to attack from diseases or pests. High nitrogen causes rapid plant growth and tender plants as well as being the main cause for many types of fungus.

6. Insects develop immunities to pesticides. While it is easy to understand that pests develop immunities to pesticides, I find it hard to believe that people think that they are immune to pesticides at all. Think about it people.

7. Avoid Killing whenever possible.   The idea that pest control means killing the pest has developed through the high pressure tactics of chemical sales people. To sell more and more chemicals is indeed their goal. Killing upsets the natural balance. Let Mother Nature do this for you. She has already set up a system of her own that works. There are many ways to control the behavior of pests. Only as a last resort do you kill pests. This is not a radically new idea. All cultures through out the world respect their insect brothers except the white man.

Making the Organic Transition

Proper nutrition is the very foundation of organic growing and pest control. Just as a human body that is not properly fed is wide open to attack from one illness or another, so it is with plants. Chemical fertilizers are not complete foods. They lack almost all of the important trace minerals. There are over 70 trace minerals that plants and humans need, and chemical fertilizers provide only a few of them. If plants do not get the trace minerals they need, as well as a variety of enzymes and bacteria, they will become stressed and will be attacked by disease, or pests.

Ants, like most predatory insects, attack plants that are sick or stressed. This is a natural law, like the wolf who only attacks sick sheep. I do not recommend the use of chemical fertilizers. If you presently are using these fertilizers and want to stop, be forewarned that you must proceed slowly and carefully as you make the transition from chemical fertilizers to purely Organics. As an addict depends on drugs, plants depend on chemical fertilizers for their 'hits' and will not accept organic food at first. Time is needed to allow the plants natural systems to begin to work again. Give yourself and your plants at least a year to go from chemicals to Organics. It may take several years of growing organically for the soil to return to life and for the organic materials to kick in. Use the following schedule in making the organic transition.

First Month: Keep using the same amount of chemical fertilizer but also spray a liquid seaweed blend (such as SuperSeaweed, or NITRON or AGRI-GROXE "Agri-Gro") as a foliar feeder. Make sure the soil is in good condition by applying lots of rich alive compost. Stop using any chemical pesticides. Start using the many methods mentioned in this book.

Second Month: Reduce the amount of chemical fertilizer by one-third, while continuing to spray the leaves of your plants with the liquid seaweed according to the instructions (Superseaweed is five drops per gallon). Replace the chemical fertilizer with a good compost such as Organa. Apply more compost around plants, turn over soil when possible. Mulch well.

Third Month: By now you should have reduced the amount of chemical fertilizer to one-half the original amount you were using before and have increased the amount of organic fertilizer to replace the chemical fertilizer. In other words, you are now using one part organic and one part chemical. You will find that during these months your plants will be under a great deal of stress and may be attacked by various predatory bugs. Use only organic methods to get rid of them. Keep spraying the leaves with a good liquid seaweed weekly or monthly.

Fourth to Eighth Month: Stay on the 50/50 basis. This is a good time to see how the plants are responding to their new organic regime. Some plants will not like it at all and may even die, but the majority of plants will do very well. Keep spraying the leaves as often as recommended (as above).

Ninth Month: Reduce the use of chemical fertilizers by one-half again, and increase your organic fertilizer/compost. Follow with a monthly spraying of liquid seaweed.

Tenth Month - Twelfth Month: Repeat the process. By now you should be using one-quarter of your chemical fertilizer with three quarters organic fertilizer. This is the level you want to stay at for the next two months. By the end of the first year you should have reduced your chemical usage by 80 percent. You should also have increased your organic usage by 80 percent. During the next few years you should work on reducing the chemicals used by ½ then eventually not using any chemicals at all and be 100% Organic! Any problems your plants may have can be treated with the appropriate organic controls; some of which are mentioned in this book.

The Invisible Gardener says:

"The best way to help mother nature is to get out of her way!"

The Invisible Gardener says:

"It's not the plants that are

hooked on the chemicals,

It's the people using them!"

---

2 Frogs actually eat up a great deal of insects and should always be welcomed in the garden.

*The Invisible Gar-
dener says:*

*"Hand-picking" is
the best!*

Using Liquid Seaweed

Liquid Seaweed is very important because it provides plants with the necessary trace minerals to build up their immune systems. It also provides bacteria for the plants' natural systems to work with. Use distilled or spring water or well water for this as city water may kill off the bacteria (or you can filter the water. See garden filter page.). Read the chapter on organic fertilizers in this book for details on what organic fertilizers to use. Remember that the first line of defense is careful planning. Many plant troubles blamed on diseases and pests are really caused by poor soil. A good organic grower plans ahead. He or she works into his soil as many different varieties of organic materials as he can get. He or she adds manure and mulches and provides the soil with as much and varied sources of nutrients as possible. This is the key to successful organic growing as well as to organic pest controls.

Proper Watering Methods

If a plant is improperly watered, it will suffer stress and open itself up to attack by insects and rodents. The lack of regular watering causes nutrients to disappear and become unavailable to plants. Learn the correct methods of watering. Too much water is just as bad as not enough. Insufficient lighting also contributes to stress. Trees weakened by improper watering, root damage and improper feeding are more likely to be attacked by pests such as borers than are healthy trees.

Know your Insects

Know your pest. Get a hand lens and examine the stems and leaves and fruit surfaces to see what is causing the damage. Look for tiny eggs, tiny puncture holes, trails, excrement and insects. In other words identify the problem. There are many good books on insect identification on the market[3]. Not all insects are bad, so be careful which ones you place the blame on. Also learn to identify insect eating habits. Some insects are chewers such as caterpillars, earwigs and leaf miners. Others are suckers such as aphids, and mites. Still others feed below ground such as nematodes, maggots and rootworms, or above ground such as apple maggots, beetles and tomato fruit worms.

Each insect requires a different method of treatment. Diseases also demand various different methods of treatment. Learning how to recognize different types of insect damage and their associated diseases takes time and experience. There are many books available that cover this subject of identification. **Rodales Organic Gardening Association** has many fine books on this subject.

Physical Barriers

Physical barriers are an important tool for the organic grower. By leaning to use the many natural barriers available through hot products such as those mentioned in the hot barriers list, you can keep pests off your plants as well as control rabbits, deer and other animals from eating your plants. Works great on below ground animals too! (See Hot Physical Barri-

---

3 A good one is Natural Insect and Disease Control by Roger Yepsen Jr (Rodale Press, 1984). Another good book is Common-Sense Pest Control by William and Helga Olkowski and Sheila Daar (Taunton Press, 1991).

ers List, index)

Here is a good formula to use for a physical barrier:

The Plant Guardian Formula

10 parts Dia-earth

1 part Pyrethrum powder

1 part Organic Tobacco[4]

2 parts Kelp Meal like Acadie

1 part African Cayenne Pepper[5]

1 part Rock Dust or Greensand

1 part Organic Alfalfa Meal

½ part Sulfur Dust

1 part Garlic Powder.

OK to use hot powders instead.   See Hot Physical Barriers list. Mix all ingredients together well. Depending on how much mixture you need, use a cup for the part. Run thru a grinder if you can. A coffee grinder works well. Wear face mask and mix slowly. Use when needed around plants. Can be used with Pest Pistol or the Duster Mizer. This is a special formula developed by The Invisible Gardener. You must make your own as this stuff is not for sale. Potent! Use as a dust around plants. This formula can also be used to create a paste (slowly add water to mixture, and stir until it turns into a paste). Paint the paste onto plants with a paint brush. It will last a long time and will protect plants from crawling insects. Retouch with paint brush when needed. Great for painting raised garden beds to protect flowers and vegetables. The paste will kill many insects that come into contact with it. The paste also will kill beneficial insects as well, so use on spot applications only. Dusting is best for crawling pests such as snails and slugs, spiders, snakes, rats, gophers (place into tunnels), ants, fleas (outdoors only), and will keep, rabbits, deer and coyotes away (sprinkle around property edge).

Tangelfoot: Made from castor oil and bees wax. Very sticky stuff. Place around trees as a barrier. (I like to mix the Tangelfoot with Plant Guardian to make an even more potent barrier). But be careful not to over do it as Tangelfoot can damage the bark of the tree. So only apply as little as needed and then remove when no longer needed. Use a tape to avoid damaging trunk.

Tabasco Sauce (see also Peppers): Makes an excellent barrier to many insects. Simply add one tablespoon of Tabasco sauce into a quart sprayer. Spray on an area you want insects to stay out of. When it dries it leaves a residue behind. Another favorite of mine is to use clay and mix it with cayenne pepper. I use it to paint the trees or dust around the outside of the garden. Snails hate this stuff! Buy only that which has no additives and preservatives and is organically grown.

The Hotter the Source- the Deader the Pest!

Hot and spicy recipes are available from around the world; a good source is a mail order company called Mo Hotta Mo Betta @ 1-800-462-3220.

---

4 Tobacco is dangerous to breathe or ingest so use with caution. I prefer the smoking or pipe tobacco instead of the dust. Try to use without preservatives etc.

5 Remember to try and get the hottest chili or pepper mix that you can.

Hot Physical Barriers List: A hot physical barrier is using a dust that when an insect walks over it, will either kill it or keep it off without hurting the plant or soil. Pepper based barriers are the easiest but many other natural hot barrier are also available. These barrier also make effective repellents against deer, rabbit, snakes (yes, snakes hate hot physical barriers).

Here is a partial listing of some of the hottest physical barriers around. These are all available from the **Mo Hotta Mo Betta Catalog,** if not available at your local store. Some come as a powder and some as a liquid. Use only products without additives or preservatives please! Best to grind.

Hot

Habanero Hot Salt , three dried peppers: Habanero, Chipote, and Tabasco

Hot Mama Rap's Hot N' spicy seasoning: garlic, chilies, spices, bell peppers, onion

Texas Gunpowder dried jalapeno powder

Crushed Chipote chile powder

Very Hot

MoHotta MoBetta hot habanero flakes

Goldwater's Senator's chili mix

Jardines Texas Chili Works. Has masa powder

Big Bruces Gunpowder Chili. Use it with the heat turned all the way up!

Very very Hot

I am On Fire- Going to Heaven 50,000 Scoville heat units

Very, Very, Very Hot!

I am on Fire- Ready to Die! 180,000K heat rating

Red Hot!

Mad Coyotes chili from hell

Traps: Japanese Beetle Traps are a pheromone (a natural insect attractant) that attracts females and traps them for disposal. Fly trap, see fly chapter. Apple maggot Traps, apple sized red spheres coated with a sticky substance. Attracts females maggot flies. Use Tangle-trap a brush on sticky substance for above trap. Homemade traps are easy to make that catch any type of insect. Experiment with using apple cider, vinegar, whisky. Sticky Traps, comes in either yellow or white. Attracts and traps them. There are many Light Traps, pheromone traps and sticky traps available for many different types of insects. A good source of these traps is ARBICO, or Gardener's Supply. Traps can be made more effective by using the various fermented products on the market such as apple cider beer (fermented apples), vinegar, wines, whisky, etc.

Biological Controls: The use of beneficial organisms offers an effective system of integrated pest management. Not all insects are bad. The good guys and the bad guys do not live in the same place. When the good guys move in, the bad guys move out. It's that simple. Some insects help to improve the soil through aeration, while others help by generating compost from leaves and other decaying organic matter. Some insects, such as bees, also help in pollination. Without the help of insects, we would be covered with our own waste. Insects recycle waste.

Biological Controls using a friendly insect to get rid of the unfriendliest should be used in conjunction with an integrated system of pest controls: monitoring, identification, trapping, planting pest resistant varieties, proper fertilization, proper irrigation and the use of natural insecticides. Care must be taken in releasing biological controls to provide them with a beneficial environment. Provide for an environment that promotes a diversity of predators and parasites. A wide range of plants also is important. Before releasing the beneficial bacteria, water the area well. This provides for immediate sources of water for them. Release beneficials in evenings only since hot days would make it difficult for them to establish themselves. They need to find a safe hiding place. Avoid using sprays, barriers when using biological controls as it will hurt them also.

Bio-Insecticides: When using BT, add a little molasses to increase effectiveness. A natural UV inhibitor will make it last longer and not break down from the sunlight. A good natural sticking agent will help here. Oils make good UV inhibitors.

BT (Bacterium Thuringiensis) discovered in 1911, is a type of natural bacteria found in the stomach of certain caterpillars. There are over 35 BT varieties on the market. Use only when necessary as pests will develop an immunity to this. Controls caterpillars, corn borers and mosquitoes. A stomach poison that stops insects from feeding and become paralyzed. Effects only targeted insects not effecting the environment.

BT, Kurstake Bait, controls European corn borers. Apply early spring.

BT/San Diego M-One, controls Colorado potato beetles.

BT/H-14 Gnatrol controls fungus gnats. Fungus gnats lay eggs in soil. They love potted soil, peat moss and organic rich humus. The larvae feed on plant roots and root hairs crippling (if not killing) the plants. Population is highest during winter month and spring. Infested plants show discoloring, wilted leaves, root rot. They attack ornamentals, many vegetable crops, poinsettia and others.

BT/Israelensis controls mosquito and black fly larvae. Adult females drop eggs in water, to hatch under water. 98% die after only 5 min. of low exposure to this product! Kills only larvae. Harmless to fish, birds, mammals, plants and the environment.

MVP (a natural bio-insecticide) is a new version of BT. Stays effective for as long as eight days. Effective against most worms such as armyworms, imported cabbageworm, cabbage webworm, cabbage looper, hornworms, corn borer, leaf rollers, moths and most caterpillars. Will not harm mammals and beneficials, fish or birds. Use on cabbage, broccoli, root crops, lettuce, corn, tomatoes, peppers, citrus fruits, peanuts, apples and many more plants. Can be applied up to day of harvest. Use 1-1/2 to 3 oz. per gallon. Available from ARBICO or through mail order catalogs.

"Biological Controls using a friendly insect to get rid of the unfriendliest should be used in conjunction with an integrated system of pest controls: monitoring, identification, trapping, planting pest resistant varieties, proper fertilization, proper irrigation and the use of natural insecticides."

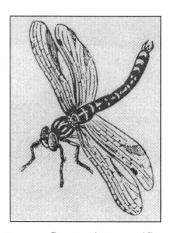

Dragonflies and Damselflies

_____

"Birds eat many different types of insects year round. Providing housing, food and water for our bird allies will help insure good insect balance."

_____

The Woodpecker

Various Varieties of Beneficial Insects

Semaspore, Nosema locustae, for control of grasshoppers. A naturally occurring protozoan is deadly to grasshoppers and crickets that eat it but is pest specific and will not harm any thing else. Infected grasshoppers spread the disease to newly hatched grasshoppers. Apply in early spring for best control.

Aphid-lions or Chrysoperia Camea and Chrysoperia Rufilabris: Some aphid-lions are also known as dobsonflies, ant-lions, or lacewings. The aphid-lion is found in most gardens and are predaceous. They are an all purpose garden predator. The larvae eats aphids, mealybugs, scales, thrips, mites, spider mites and whiteflies. They destroy many other destructive insects, as well the eggs of many caterpillars, mites, scales, aphids and mealybugs.

Lady Beetle or Ladybug: Many varieties are native to the United States. Both the young and adult stages eat various soft bodied insects such as aphids. If a good food source is available, the lady beetles will stay and lay eggs.

Dragonflies and Damselflies: The Mosquito Hawk is a good name for the dragonfly. They have highly developed eyes and a speedy mode of flight. They also are fierce hunters. The damselfly is the smaller of the two and unlike the dragonfly, folds her wings on her back. The young of both the dragonfly and the damselfly are called nymphs or naiads and they devour mosquitoes and other water-born insects.

Fly Parasites: Attack flies before they hatch with this parasite. Fly parasites deposit their eggs inside immature fly pupae, the parasitic eggs hatch into larvae which feed on their hosts.

Praying Mantis: The Chinese mantis was first introduced into the United States in 1896. The mantis captures, holds and devours many different bothersome insects. A very helpful insect in the vegetable garden or around the house.

Spined Soldier Bug: Controls Mexican bean beetle: This bug preys on many garden pests such as cabbage loopers, cabbageworms and Mexican beetles. Both adults and nymphs attack other pests.

Semaspore: A predator of a different sort. A long term grasshopper control which is deadly to grasshoppers and crickets, is made from a naturally occurring protozoan.

Other Predators: The most abundant and important predators for a healthy garden are the two-winged Flies, the Wasps and the Green Lacewings. Parasites will attack insects in all stages of their development. The host does not die immediately but provides nourishment until the larva of the parasite is nearly grown after which the host dies.

Tachinid Flies: These predators prey on a wide variety of insect species. They lay their eggs in the host body, which provides the young with a source of food. Compsilura concinnata is a fly that was imported from Europe to combat the gypsy moth. Flesh flies have varied habits: Some are flesh eaters, while others eat only insects.

Wasps: Wasps feed mostly on other insects. Their favorite is caterpillars such as the armyworm. There are many varieties of parasitic wasps. Wasps are generally quite social, with males, females and sterile workers making up the family. Some species of parasitic wasps attack only certain insects. Encarsia Formosa is the whitefly parasitoid which effectively controls whiteflies.

Lysiphlebus Testaceipes: These bugs destroys millions of aphids per year by laying eggs within the bodies of the aphids.

Trichogramma Minutum and T Pretiosum: These are a minute egg parasite that destroy the eggs of most injurious pests such as the bollworm, cotton leafworm and various borers, hornworm, codling moth, and all moths and butterfly eggs.

Honey Bees: Honey Bees are very valuable in their role as pollinators as well as a source of food.

Beneficial Nematodes: These parasites control many borers, grubs, cutworms, oriental beetles, pillbugs and cutworms. Available are parasitic nematodes of the HB and SC varieties. The HB variety is an effective alternative to milky spore for the Japanese beetle grub. Medfly larvae can be controlled by these also. See Bio-Halt!

Clandosan: These parasites control root-knot nematodes, and simulate naturally occurring soil organisms. They should be tilled into the soil before planting, and then applied every year after that. They also can be used as a side dressing as needed.

Fire Ant Bait: Contains avermectin which is a natural soil organism. Ants take bait back to colony where it is distributed through out colony and kills entire colony within 3 months. Use the same system as on ants.

ARBICO and/or Rincon-Vitova (both IGA commercial members), are good sources of biological controls, predators and parasites.

For the Birds: Birds eat many different types of insects year round. Providing housing, food and water for our bird allies will help insure good insect balance.

A few examples: **The Woodpecker:** Loves wood boring larvae of beetles, ants, borers and many other insects and their larvae, **The Chickadee:** Loves lice, caterpillars, flies and tree hoppers, **The Bluebird:** Eats sowbugs, caterpillars and other insects. **Owls** eat at night many different types of insects. **Mockingbirds:** Will eat termites, caterpillars and beetles, **Robins:** Also eat termites, they love caterpillars and beetles, **The Cardinal:** Likes grasshoppers, ants and beetles. Can You imagine how many insects baby birds eat?

Bats: Bats eat a great deal of insects in their life time. Protect them when ever you can. Bat houses are available through many sources try Real Goods (an IGA member since 1997).

Vegetable Organics

Vegetable Organics is not a new science since housewives have been using garlic or cayenne pepper to control bugs for centuries. Veganics is also a safe method of using not only the wide array of materials available at the grocery store, but also the wide array of vegetables that we can grow ourselves! From pepper to lettuce, there are many store bought items that

"Natural Insecticides made from the Earth, return to the Earth"

It is to our advantage to let Nature be our teacher and help us learn from our mistakes. Better Living Through Chemistry[6] has helped us to develop into a highly skilled and educated society. Better Living Through Nature can take us even further by enabling our societies to grow and develop sustainably while we exist in harmony with Nature.

6 Remember that old slogan?

can be used safety against pests[7] .

Homemade Sprays (see also Hot Sauces, Tabasco, Chili Peppers): Homemade sprays can be made from a variety of plants, herbs, vegetables and various other bought items. Always use with caution. Follow the rule "Less is Best". Use as little as you can to achieve the results you want. There are many store items that are essentially homemade receipts gone mass market. The craze for hotter chili sauces and hotter sauces of any kind has made it easier for the organic gardener to find many alternatives to pesticides for his pest control. The hotter the better! Avoid anything that has preservatives or additives, colorings, etc. A good mail order source is **Mo Hotta Mo Betta.**

A synergistic plant is any type of plant or natural material that when added to another plant and or material increases the effectiveness of that plant. **Soap** is a natural synergist for most plants as it acts both as a wetting agent and pest and or disease control. Tabasco sauce is another such product that kills on contact and repels as well. **Tabasco soap** is a blend of the two. Many vegetables can be made into a liquid and then sprayed in a diluted form to protect plants, repel or even kill bugs. There are many bought items and vegetables that are all ready made for you to use such as those mentioned in **Mo Hotta Mo Betta** catalog. Some good oil synergists are garlic, peanut, pecan, coconut oil, any vegetable oil. Use all plant mixtures with caution.

There are many plants that will kill both beneficials and the pests as well, so be specific as to what you are applying or spraying. I have found that almost all plants can be harmful if improperly introduced into the garden environment. Tobacco can create havoc in the garden if sprayed without proper care. Cayenne pepper does the same damage also. It is for this reason that you must use with care any sprays that you make. Also many plants can be harmful to humans, birds, fish. Because it is labeled organic, it does not mean that it is safe to use. One drop of Tobacco Sulfate is enough to kill you. As simple a plant as garlic can do a great deal of damage in the garden's eco-system as it affects the beneficials as well. Many plants are oil based and may produce toxic substances as well. The Oleander plant gives off a toxic substance that could kill a young child or dog if enough of the plant is eaten. Lettuce when made into a liquid will kill white fly and many other soft bodied insects! Leaves from your tomato plant will kill roaches and flies! Soap if over used will kill beneficials and earthworms and make for a dead soil. So be careful with these mixtures. Label them and teach yourself, your children how to use them!

Wash off all fruit and vegetables, before eating!

Types of Preparations

Infusion: You can make an infusion from almost any plant, vegetable or herb, leaves and flowers. An infusion is a fast and simple way to get immediate relief from pest or disease attacks on your plants. This often times will provide pest and disease protection and you will not need to make any thing stronger. Learn to blend different plants to achieve best results. Pour boiling water on to the leaves, flowers, roots, seeds and place in a cooking pot or container with a lid. Allow to steep for 1 hour. Stirring the mixture regularly. Pour through a strainer. Label for use later. Can be bottled. Store in cool dark place. Strain before use. Use ½ to 1 ounce in a pint of water or can be used straight. For best results do not boil but bring water to a boil then remove from heat. Always use glass, porcelain or enamel cooking utensils if possible. Can be sprayed onto bugs or plant while it is still hot! It will not burn plants but will kill insects. Best for diseases. Used this at full strength to control diseases.

Decoction: This is made by simmering the part of the plant (usually roots, twigs and bark) in a nonmetal container, in boiling water, for ½ hr or more depending on the material(s) and its use. Always start out with more water then you need to allow for evaporation. Boil until water is dark. Strain before use. Always use filtered water if possible. This method is used for extracting the active ingredients. Store in a dark colored glass. Label for later use. Makes a much more concentrated mixture. This mixture is made when a strong mixture is needed to provide a sure kill or control of the pest or disease in mind. Great care must be taken when using not to upset the eco-balance. This mixture will kill beneficials as well as pests. Many store bought items fall into this category. A strong formula is great to kill off pests and control and diseases. Can be used when it is hot. Use at full strength.

Pain Away: This is a blend of oils that includes Helichrysum. Birch, Clove and Peppermint. It was developed to help reduce inflammation, promote healthy circulation and healing, and alleviate pain and because of its strong odors will repel most insects as well as keep them from eating your plants. Ants hate this stuff! Helichrysum is another oil with a strong smell to it. This smell will repel many insects. Clove is one of my favorites to use. Clove stops most insects from eating. Think of using at drops per gallon.

For more info see specific plant or herb.

PLANTS TO USE FOR NATURAL PEST CONTROL

Most of these you can grow yourself! We are constantly expanding this list with every printing. You can make the preparation suggested or you can obtain it from the herbal store. Everything mentioned here has been used by The Invisible Gardener in his services.

Anise [8]: To repel pests make an infusion, allow to sit for an hour add a dash of Tabasco and Dr. Bronners Peppermint soap to increase effectiveness. Discourages most chewing insects and caterpillars. Decoctions and Tinctures provide most protection to vegetable plants. Extracts should be used with care as it will repel beneficials also. A good synergist. Easier to buy anise oil.

Sugar Apple[9]: Use the seeds, leaves and roots[10] to make an extract. Highly toxic to most insects. Dry and grind to a powder. Add ½ cup to 4 cups boiling water. Allow to sit overnight. Strain.

Azalea: Dried flowers can be used as a dust, or infusion. A contact and stomach poison. Avoid use on vegetables.

---

7 When buying just follow the simple rule of no additives or preservatives unless natural.

8 Avoid using on fruits and vegetables unless you wash off well.

9 Avoid use on fruits and vegetables otherwise wash off well.

10 The roots being the strongest part of this plant. See appendix for prep of roots.

Grinders: Grinders are useful tools in preparing your natural barriers, are for grinding down the various parts of the plants from seeds to peppers, peppercorns, etc. European Steel Mill Peppercorn Grinder, adjustable is a good one. Available from Mo Hotta Mo Better Catalog, Real Goods, Gardeners Supply.. Magnum Peppermill is an excellent small tool. The Peppergun is an excellent product for grinding many things. Also available from Mo Hotta Mo Betta. Use the grinder to grind the peppercorn directly onto the plants and as a barrier. Avoid watering for 24 hours for best results.

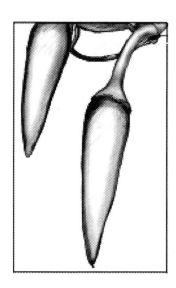

Cayenne Pepper

Make solar tea from dried leaves and flowers (see appendix on preparing solar teas). Add one cup of dried leaves or flowers into the leg of some panty hose and place into a one-gallon glass container. Add distilled or filtered or spring water, and one tablespoon melted cocoa butter, coconut oil (or the tan lotion mentioned), per gallon of solar tea. Add one cup NITRON A-35 and one cup AGRI-GRO or any liquid seaweed per gallon. Add one drop natural soap (such as Dr. Bronners Peppermint soap), depending on the insect. Add a dash of Tabasco sauce to increase effectiveness. Test first for strength.

Balm[11]**:** When used will repel aphids, ants, and most insects since balm have few pests that like to eat it. Balm oil and Peppermint oil make a powerful repellent. A good synergist. Balm is easy to get at your drug store, etc. Add a capful per quart water.

Basil or Sweet Basil Oil: Is very affective against many insects from mosquito larvae to house flies. An affective synergist to pyrethrum and tobacco. Use 1 oz[12] per gallon for infusion. Experiment. Extracts are very powerful. Add a dash of Tabasco and soap. Also effective as an infusion, decoction. Also sold as an essential oil. Effective against borers, and most caterpillars. Use with a little Tabasco and soap. Make an infusion. Essential Oil is also available. Avoid on fruits and vegetables.

Beer: Beer is produced from fermented hops. Fermented products are treated differently then you might think by the plant kingdom. It is the fermentation process that converts energy to the plants as readily available sugars, and minerals. I would suggest that you experiment with the strength since lawns require a different strength then say roses or your vegetable garden. The main problem I have with most beers is the chemicals that are in most mass produced beers. There are over 72 different chemicals in beer! From pesticides to colors to anti foaming to foaming agents. Nothing organic here! So I would suggest that you make it yourself for your own use and make some extra for your plants!

Beet Juice: When mixed with Tabasco sauce will effectively repel most flying insects. Controls many diseases. Make from a juicer.

Borage Oil: An infusion will control leaf eating insects and will also repel.

Cocoa Butter or Coconut Oil makes a perfect synergist for many plants. You can also use sun tan oil made from cocoa butter[13] . Use only a small amount to start. Try one tablespoon per quart. Experiment! These oils will kill many different types of insects both pests and beneficials so be careful when using! Will kill soft-bodied insects.

Cabbage: Leaves can be used to attract aphids to traps. Make an infusion. Add a dash of Tabasco (or other hot sauces) soap. Allow leaves to simmer in water. Use at full strength.

Caraway: Creates a strong repellent against chewing insects. An infusion with a dash of Tabasco soap will protect most plants. Test for strength.

Castor Oil: The Castor Beans produce Castor Oil, makes a perfect synergist for pyrethrum and other natural insecticides. Make alcohol concentrate from crushed beans or buy concentrated oil. Use 1 oz per gallon for most uses. Do not use on vegetables or fruits!

Catnip: Prevents insects from establishing on plants. Prevents worms or caterpillars if sprayed regularly. Make an infusion. Test for strength.

Cayenne Pepper (see also peppers, hot sauces): Will destroy many insects that are either dusted or sprayed. A good source is Tabasco Sauce made from cayenne pepper, vinegar and a dash of salt. Really great. Read the ingredients to make sure that there are no additives or preservatives. All parts of the plant can be used. The powder can be sprinkled around plants. The pepper can be made into an extract. Experiment with different types of peppers. Cayenne Pepper Tea?

Chamomile: An extract will repel beetles and chewing insects. Add a dash of Tabasco soap. Make a tea for yourself then a second batch for the pests! Easy to make as a tea. Infusion is best. Works against most diseases as well.

Chile: (see peppers)

Chives: Use with a natural soap to repel insects. Chive tea any one? Make an infusion.

Citronella Oil: Used for centuries as a mosquito repellent, can also be used as a synergist for many natural sprays. Avoid use on Oil based trees, etc., as pine trees. Natrapel repellent is made from citronella oil and aloe. Repels many insects. Buy essential oil for this.

Citrus Oil: Contains Limonene. An extract made from citrus peels. Makes an excellent synergist to many natural sprays. Will effectively kill many soft body insects and repels many others. Use only on effected areas. Use not more than 10 drops per gallon of the extract, depending on the pest. Use leaves, flowers, grinds. See formulas for strength. Use at weak strength for best results and less damage (soap will burn plants). Many companies on the market carry citrus products such as soaps. Citrus or lime juice can be sprayed on most plants and vegetables. Will control fleas, larvae, and most chewing insects. Citrus soap makes an excellent synergist for most extracts. There are citrus extracts available on the markets today that are safe to use. Try some of the air fresheners made from citrus. Use only a few drops per gallon. Citrus Oils are powerful extracts, used with care as it will kill off beneficials as well. **Citra Solv** is a citrus concentrate available at most stores. **Jungle Rain** is made from Citrus Oils and Peppermint soap. Another great product!

Coconut Oil: Many parts of this palm can be used. From the leaves to it's sap. The easiest part to use is what comes from the oil inside the nut. This oil is found in many suntan lotions. Read the label. Buy only 100% pure. Coconut oil and soap (coconut oil soap is the best) are a great combination. This oil is safe for humans to use but be careful to use only on specific areas as this oil will kill many insects. Can be used as an excellent synergist to be used against hard shelled insects such as snails, beetles, etc. Extracts can be made using a slow heat method.

Coffee Arabica L: Caffeine has several strong chemical compounds that are insecticidal. Soap and Tabasco sauce make an excellent synergist for this plant. Add one to 10 drops per gallon[14]. Use only natural soaps like Dr. Bronners, Citrus Soap, Herbal soaps, etc. DE also makes an excellent synergist for coffee.

Compositae: Thistle or Aster Family: Chrysanthemum cinerariaefolium. Is the plant's flower that produces pyrethrum. The flowers are dried and the powder is used to make an infusion. To this family

---

11 Avoid use on vegetables

12 See appendix for making infusions. 1 oz of Basil Oil per gallon is very strong and effective against most flying insects. Experiment with the strength. Use less whenever possible.

13 Like Hawaiian Tropic which is made from mineral oil, coconut oil, cocoa butter, aloe, lanolin, eucalyptus oil, plumeria oil, mango oil, guava oil, papaya oil, passion fruit oil, taro oil and kukui oil. Wow!

14 The strength you use depends on the type of plants you are spraying

---

You can make solar tea out of Peppermint and Spearmint, and herb teas such as Lemon grass, Citrus and Lipton tea. Start by placing 1 cup dried herb into panty hose. Tie into a ball. Place into 1 gallon glass container. Add 10 drops Superseaweed or ½ cup seaweed powder per gallon and 1 drop natural soap and a dash of Tabasco sauce or any other of the hot sauces mentioned in this book. Place in sun for two to three hours. Strain. Add to sprayer and use on plants. Experiment with various strengths, length of time in sun, etc. Experiment with different types of herb's and other hot products and keep notes on the effectiveness of your different mixtures. If you find something that works better than anything mentioned here, please send it to me for possible inclusion in the next revision. Wash off all fruits and vegetables before eating!

Give Pests a Coffee Break?

Method #1

Save the unused coffee in a glass gallon container. Use as needed. Can also be fed directly to plants. Turn over soil. Can be left as a mulch to control most crawling insects such as snails, slugs, caterpillars.

Method #2

An effective way to control most insects and also help to provide nutrition is to use your coffee grinds and to make a coffee solar 'brew'. This is done by dumping a cup of this morning's coffee grinds into a panty hose. Place it inside a one-gallon wide-mouth glass container of water and allow to sit for 24 hours in the sun. The next day, add one cup per gallon of NITRON A35 or AGRI-GRO or ROOTS PLUS[15]. Also add one drop of any of the natural soaps mentioned. Add 10 drops Superseaweed per gallon, and one tablespoonful of any kind of rock dust plus a dash of Tabasco sauce. This mixture will reduce insects and help to control many more. The mixture should be stirred clockwise for five minutes then counterclockwise for another five minute before it is strained.

belongs Dahlia, Coreopsis, Marigolds, Aster, Cosmos, and many others. Harmless to mammals.

Coriander: Makes an excellent repellent. Stops most chewing insects. Seeds are ground into a powder. Or you can make an infusion from the plant.

Cucumber: Very strong repellent of worms, ants, fleas, beetles. Seeds are the strongest. Grind into a powder or use as an infusion[16].

Curry: Made into a paste painted on trunk of plant to protect from insects and can also be made into a liquid and sprayed. Run through blender till liquefied. Strain if you need to. Try 1 tablespoon per gallon water. Test for strength.

Dill: An extract (Dill Oil) made from dill will repel most flying insects. Interferes with receiving signals.

Eggplant: A part of the night shade family. Make an infusion for best results. Eggplant oil is good also.

Eucalyptus Oil: Makes an excellent synergist for many natural sprays plus will kill on contact many soft bodied insects as well as repel them. Try Dr. Bronner's Eucalyptus soap.

Fennel: Stops all chewing insects if used as an extract. 1 drop per gallon plus a dash of Tabasco soap.

Firemist Spray: A Hot sauce not a pepper spray - the only spray on hot sauce on the market is Medium to Hot. Works great on many insects. Ants will avoid for days! Available from Mo Hotta Mo Betta catalog or at local store. Test for strength.

Garlic: You can use all parts of the plant. Use the greens, run through a juicer. Add 1 cup per gallon, allow to sit 1 hr before use. Do not allow to sit longer. Garlic Juice made from garlic cloves is stronger. Extracts are readily available or you can make your own. Use 1 oz per gallon liquid for most insects. Use soap as synergist. Hot Garlic Oil is excellent to use. Garlic Oils are available at your local store.

Researchers at the University of Washington in Seattle have saved thousands of trees from destruction by deer and rabbits by planting pellets beside them that give off the smell of garlic. The pellets are gradually dissolved by rainwater, absorbed by the roots and then spread throughout the entire tree. The tree then eliminates the compound as dimethyl selenide (the smell of garlic breath). The deer and the rabbits find the smell offensive.

Ginger: You can buy Stir Fry Spicy Ginger which has garlic, sesame seeds, ginger, onion, celery, green and red chili peppers, coriander, lemon and orange oils, chives, star anise, cloves, cinnamon, fennel, black pepper and sechuan pepper. If not available in stores try **Mo Hotta Mo Betta**. Experiment for strength.

Gourd Family: Gourd melon, squash, and pumpkin all in same family. Seeds are crushed into oil or dust concentrate. Leaves can also be dried and used as an infusion, oil extracts can be also sprayed. Will repel most chewing insects.

Hops: Make an infusion. See also beer

Horseradish: A very strong plant. Most insects will not eat this plant or have anything to do with it! I suggest getting an extract. From the roots. A good concentrate is called Extra Hot Horseradish available in stores or through Mo Hotta Mo Betta.

Hyssop: Hyssop Flowers has been said to be as strong as pyrethrum flowers. Use an infusion first if not strong enough make into a tincture.

Larkspur: Powdered roots are toxic to most insects especially caterpillars. Make a decoction.

Lettuce leaves: When made into an infusion will repel aphids and whiteflies. Add a dash of Tabasco soap. Wash before eating.

Marigold Flowers: Use dried leaves for an infusion spray. Add a dash of Tabasco (or other hot sauces) soap. Will prevent most insects, caterpillars from chewing on leaves of ornamentals.

Marjoram: Most herb's can be made into either an infusion or extract and sprayed. Will repel most insects.

Mustard: Some products on the market with mustard are:

- ◆ Orange and Hot Mustard (Hot)
- ◆ Blue Crab Bay Horseradish Mustard (Hot)
- ◆ Red Chili Peppers and Garlic Mustard (Very Hot)
- ◆ Hot Raspberry Mustard (Hot)

Mustard can be used in many ways. **For Pest control:** add 1 tablespoon mustard (or any of the ones mentioned above or substitute your own) to 1 gallon water. Stir in well. Allow to settle. Strain water into gallon sprayer. Add a dash of natural soap as a wetting agent and to enhance its effect. Or can be painted on Trunk to protect against crawling insects. Repels aphids, whiteflies, mealybugs, deer, rabbits.

For Diseases: Can be painted around bases of fruit trees, roses, etc. to keep ants etc., off. This will allow disease to be controlled better by controlling vector insects. An Infusion can be made from the mustard plants themselves.

Nasturtiums: An extract can be made. Sprayed on plants will repel aphids. An infusion of the plants can be made also.

Neem Tree Tea Oil: Excellent anti fungus, can repel many insects and destroy others. Use only on effected areas. Will also repel and kill most insects it comes in contact with including benificials. Essential Oils are easy to find. Avoid use on vegetables. Works great on most insects and plant diseases.

Nettles: Use both roots and dried leaves as well as flowers, same as above. High in calcium so OK to use for diseases. Make an infusion from the leaves.

---

15 Or any natural seaweed product,

16 Place several chopped cucumbers in a pan of water, bring water to boil and allow to simmer for 1 hour at low heat. Strain into container. Use at full strength. Sprayed directly onto bugs.

Method #3

Make a batch of your favorite instant coffee formula, add one drop of bio-degradable soap per cup and a dash of Hot Sauce per cup. Make the mixture as strong as needed, but be careful not to put too much soap in as it could burn the plants. A good rule of thumb for spraying is to spray less. Learn how to use soap. See index on soap. Add ½ tablespoon seaweed (like Acadie) or 1 capful Superseaweed per cup of instant coffee, OK to spray while hot! Can be sprayed hot directly on my soft bodied insects, as well as snails, caterpillars. Will not damage plants. Try it!

*Very Hot Curries
Asian curry- green, matsaman, Panama. red, yellow*

Indian curry- sold as a paste

Caribbean curry- Hot Jamaican curry powder

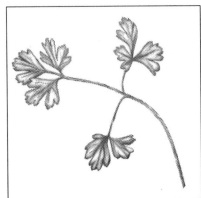

Parsley: Spray on most plants.

Here is a partial listing of Chile peppers from around the world and their heat scale:

<u>*Mild*</u>

Anaheim from Mexico

<u>*Mildly hot*</u>

Aji Limon Peruvian pepper

Aji Orchid Brazil

Poblano from Mexico

New Mexican

<u>*Hot*</u>

Ancho's Mexico medium

Cascabel

Pasilla Negro

Macho Peguin

Ricotillo from Puerto Rico and Bahamas (small light green flying saucer shaped)

<u>*Very hot*</u>

African Devil Tanzania

Datil from Florida (hotter than Tabasco)

Nutmeg Oil: Used as a synergist. Avoid using on fruits and vegetables.

Onion family: Makes a strong concentrate, use dried onions or liquid prep. Just like making soup. Allow to cool. Add a dash of Tabasco (or other hot sauce). Experiment with different amounts.

Oregano: Make an infusion. Will repel most insects. Add a dash of Tabasco soap.

Parsley: When an infusion is made, use right away. Spray on most plants. Add a dash of soap to increase effectiveness.

Pea Leaves: Can be made into an infusion that will repel most chewing insects.

Hot Peanut Sauce: From Thailand. Hot peanut Sauce is available from stores or from **Mo Hotta Mo Betta**. Very Hot. You will need to run through a strainer before using. Repels rabbits, gophers, deer.

Hot Pecan Oil: Paula's Hot-Hot Pecan Oil. Red pepper, black peppercorns, garlic. Use a screen or strain before use. Very Hot. Test for strength. Try using 1 tablespoon per gallon water.

PennyRoyal Oil: An excellent synergist for many natural sprays. Will confuse insects and/or repel and or kill. Some insects are destroyed on contact. Try one drop per gallon for most insects. Sold as an essential oil.

Peppermint Oil: Makes an excellent synergist for many natural sprays, will confuse many insects. A good source is Dr. Bronners Peppermint soap. Contains peppermint oil as a base. This oil will kill, repel or control all types of insects as well as beneficials so be careful. Will kill many insects on contact including benificials. Jungle Rain is part peppermint oil. Try using 5 tablespoons per gallon. Test for strength.

Peppers: Peppers provide the organic gardener with one of his greatest resources for organic pest control. From a dust to a liquid that you can spray. Make a decoction. Sprayed on plants or insects.

Chile Peppers

Chilies are a great source of controlling many insects either as a powder or as a concentrate that can be added to water and sprayed.

Chile Paste is an Asian hot item and is useful in controlling many insects.

Chile Pepper Oil is another Asian food item. This one is very hot and works for a longer time. Test for strength. Try 5 drops per gallon water.

Peppercorns: Peppercorns can be fed directly into gopher tunnels or ground up and dusted on plants as a barrier. Here are a few available from Mo Hotta Mo Betta.

- Malabar Black Pepper from Malabar India

- Santa Fe Pepper from Santa Fe New Mexico

♦ Pepper Royale ~ a mix of Far East, Brazil and France.

♦ Garlic Pepper ~ Black and White Peppercorns and chopped garlic

♦ Montok White Pepper ~ White Peppercorn from Bangka coast of Sumatra

Pine Oil: Kills ants when mixed at 1 drop per quart water. Use only on ants, do not spray plants directly. Buy only pure pine oil.

Potato: A member of the night shade family; extract will control beetles, and most chewing insects. Add a dash of the Tabasco Soap. Like making soup. Potato soup?

Pyrethrum: Made from the pyrethrum flower[17]. A very strong insecticide that effects both bad and good guys. So use on specific plants or pests and not garden wide. Can be used to control caterpillars, beetles, ants, aphids, mites, thrips, moths. Pyrethrum is mentioned through out this book for its specific uses. Use only as a last resort or under controlled conditions. See index for more info. See Chart.

Quassia: Bark and wood chips contain insecticidal properties. An extract will control many insects. Avoid use on vegetables or fruits. Allow chips to simmer for 4 hours over low heat. Add a dash of Tabasco soap to increase effectiveness.

Radish: An infusion of the leaves when added to a dash of Tabasco soap will repel most chewing insects, whiteflies, ants and some animals. Radish oil is best. Make an infusion of the leaves or a decoction of the roots.

Rosemary: Use roots, leaves, flowers, use same as above. An infusion of leaves makes an excellent repellent of many insects. Add a dash of any natural soap.

Rotenone: A better known plant made from the roots of a Tropical Derris Tree and from the cube root. It has been used world wide for many pests for many centuries. Very low toxicity to man and animals. It is a contact stomach poison. It is very effective against the following pests: all soft bodied insects, various spiders, spider mites, flies, snails, fleas, ticks, beetles, leaf roller, borers, etc. Sold in 1% and 5% solutions. Available from Arbico, Peaceful Valley, and most mail order companies. Rotenone/ Copper mixtures provide disease controls from insects while controlling the insects too. Dangerous to fish.

Ryania: A botanically derived natural insecticide safe to use for mammals, birds, etc. Made from the Ryania plant found in Trinidad and is in the same family as tobacco. Ryania effects insects eating and insects starve. Used as an extract will control and repel chewing caterpillars, moths, spider mites, most chewing insects, ants, flea beetles, leafhoppers, cockroaches, aphids, silverfish, spiders, thrips, whiteflies, Japanese beetles and more. Add a dash of any natural soap to increase effectiveness. Can be mixed with other extracts.

---

17 If you want to make your own get Chrysanthemum cineraiifolium seeds, grow it then use the flowers by grinding up when dry and adding to water. Allow to sit over night or make an infusion.

Very hot

New Mexican Red Chile pods

Thai from India

Habanero: Habanero Drops are great anti gophers!

Santara

Chipote Jalapeno

Afghan

Mildly Hot Red Fresno (little red pointed chili).

Serrano (long red pointed chili pepper)

Cayenne chili peppers from Africa and Asia (Most Tabasco sauces are made from this pepper).

Scotch Bonnet (Yellow flying saucer like shaped) Make into a decoction, strain and spray on plants or insects. Try 1 tablespoon per gallon.

DON'T

PANIC

IT'S

ORGANIC!

Pennyroyal Oil: Some insects are destroyed on contact.

Rue: Here is another powerful herb that when made into an extract will provide protect from most chewing insects. Add a dash of natural soap.

Sabadilla: A member of the lily family. The seeds are highly effective when ground up. A contact and stomach poison. The alkaloids in Sabadilla effect the nervous systems of insects. Highly biodegradable leaving no harmful residues in the environment. To Control aphids, blister beetles, cabbage worms, citrus thrips, German cockroaches, grasshoppers, chinch bugs. Add a dash of Tabasco Soap. Will also kill many beneficials as bees so use carefully.

Sage: The dried leaves of this plant can be used to make a solar tea. Add one cup of dried leaves to a panty hose and place into a gallon glass container filled with filtered, well, or spring water. Allow it to sit in sun for 24 hours before adding one drop of soap per gallon or you can make an infusion of leaves. Active principles of the plant are the alkaloids contained in the roots and leaves as well as in the seeds. Moderately toxic to most insects but safe to use. A stomach poison for insects causing them to stop feeding after ingesting. Effective in hot weather. Can be combined with pyrethrum and rotenone for maximum. Effectiveness. Controls many harder to control pests such as citrus thrips, corn borers, oriental fruit worm, corn earworm, codling moths.

Sesame Seed Oil: Can be used as a synergist. Use 1 oz per gallon liquid used. Try Sesame seed oil and Cayenne pepper mix available in food stores. A Hot Sesame Oil is available from Mo Hotta Mo Betta.

Spurge Family: Croton tiglium contains Croton oil used in China. The seeds are crushed into oil. Croton resin is more toxic than rotenone to insects.

Tansy: An infusion of this herb with a dash of natural soap will deter most insects.

Teas: Many teas contain tannic acid that deters insects from eating. Use with a dash of natural soap. Try Lipton Tea or Celestial Seasonings. You can use your own home grown herbs.

Thyme: Another herb that takes well to infusion but extracts is more powerful. Sprayed on plants will control chewing insects.

Sage: Controls many harder to control pests such as citrus thrips, corn borers, oriental fruit worm, corn earworm, codling moths.

Tobacco: Nicotiana spp. Tobacco: Active ingredient Nicotine Alkaloid, Nornicotine. Sold as Nicotine Sulfate in liquid form. Very Toxic to mammals. A Contact poison and a fumigant. Nicotine Sulfate biodegrades rapidly, acts extremely fast and no insect is immune to it. Be careful when using. Use only small amounts. Use Nicotine Sulfate only if you are a professional and know how to handle deadly chemicals. A safe source is smoking tobacco. Buried at base of plants. See TGM+ in index. The best to use is Organically grown without chemical additives. You can buy cigars and crumble them.

Methods of Application (see also Tree vents)

As a Liquid or as a dust or powder or as a concentrate (Black Leaf). Making a solar tea from 1 cup dried leaves is a safer way to use tobacco. Add 1 cup tobacco leaves into a panty hose, tight into knot and suspend into gallon of filtered water. Allow to sit for 24 hr. Spray on plants or pests. Allow 24 hr before harvesting. Add 2 tablespoonful of any biodegradable soap (coconut oil increases effectiveness of tobacco). Vegetable Oil can also used at 4 tsp. per gallon of tobacco solar tea made.

Tomato Plant: A member of the nightshade family, leaves can be dried, made into a tea and sprayed for many different types of pests. An infusion will control many chewing insects. An extract is more effective.

TGM+ (Tobacco, Garlic, Manure and Rock Dust): Can also be buried under plants, best added to tree vents. Plants absorb nicotine/garlic which kills any pests attacking it. Not harmful to plants, trees, roses, ornamentals. Do not use on fruit trees during fruiting stages. See tree chapter for more info. A very powerful tool!

Wild Flowers: Most wild flowers can be made into an infusion or an extract. Experiment.

Wormwood: Powerful when made into an extract. Be careful with its use. Controls most chewing insects, snails and slugs.

Using Soap

Pure soap is the most common ingredient found amongst organic Gardeners list of safe sprays to use. Most oils come from fats and oils found in animals[18], some come from plants and others from trees (coconut). It is useful against many insects as well as being a wetting agent. The most famous soap is Safer Insecticidal soap (see below) which can be found in many nurseries. Safer soap is made from fatty acids found in animals. See Soap in index for more references. Dr. Bronners Soaps has been around for a long time also and is an excellent natural soap to use. Try using soap at 1 tablespoon per gallon water. The strength will depend on the type of soap you are using, what it is made of, the time of day, the insect, etc. Always use less if possible.

Safer Soap: Controls most soft bodied insects, the soap must come into contact and therefore called a contact spray. Safer Insecticidal soap also comes with pyrethrum and contains no Piperonyl Butoxide. Controls aphids, thrips, mealybugs, spider mites, whitefly. Safer soap also comes with citrus aromatics. Another excellent soap to use is Jungle Rain. Will kill most insects on contact. Since it is a concentrate use small test amounts first. Completely safe and biodegradable. My favorite soap to use is Dr. Bronners Peppermint soap, available at most health food stores. A natural soap made from fatty acids and peppermint oil. When using soap be careful not to use too much as soap will burn the plants. It is always wise to test the soap out on a small section of the plant to make sure it will not harm the plants.

Hints on buying soap

It is wise not to buy soaps that have a variety of unnatural additives, colorings and other 'inert' ingredients. Do not use dish washing liquids! They are worse than most chemicals! Here is a list of some of the more well

Rosemary: An infusion of leaves makes an excellent repellent of many insects. Add a dash of any natural soap.

---

18 Vegetable and plant derivations are more effective and environmentally sound then from petroleum based.

known brands of safe organic soaps available which can be converted for our use as a natural part of our program: Safer Insecticidal soaps, Pure Castile soap, Vegetable glycerin soap, Pure Coconut Oil soap, Pure Olive Oil soap, Dr. Bronners soaps, Pure Herbal soaps, Boraxo Pure soap, Amway's LOC, Citrus Soap, Jungle Rain.

Do-it-Yourself You can make your own Natural liquid soap:

1..Boil 1 gallon water.

2..Stir in 1 cup instant Lipton tea.

3..Stir in 1 cup instant coffee.

4..Stir in 1 cup Dia-earth.

5..Add 5 drops of any natural scent that you like such as peppermint, or citronella oil.

6..Add 1 cup soap flakes made from your favorite natural bar soap.

7..Heat over low heat till boiling.

8..Lower heat and simmer for 15 minutes.

9...Pour into containers

10...Allow to cool.

This mixture can be used whenever you need to spray soap. Slice a strip of the soap and add to water and dissolve. Use approx. enough soap to effectively either kill the pest or effectively repel it. Remember, that too much soap will also damage the plants and the soil, so determine the correct amount that you will need for the job. Use a small amount on a test plant then record your results for later use. Nutritional deficiencies are generally the causes for the various problems that attack plants.

Making your own Tabasco Soap

Purchase your favorite natural soap. I suggest Dr. Bronners Peppermint soap as the most effective to use and also get a bottle of Tabasco sauce (without additives etc. or your favorite hot substitute, see also Hot Sauces) A good mixture is 8 oz of Dr. Bronners Peppermint to 1 oz Hot sauce. This mixture will tend to separate so stir well before using. Label it and store in cool place. Experiment with other soaps and hot sauces for best combination.

Jungle Rain: Natural Organic Soap Based Foliage Cleaner. Developed by Chris Klein who is an organic gardener in Encinitas California. This product is an environmentaly friendly organic foliage cleaner that removes mildew, fungus, spores, black sooty mold, insects and chemical residue. Non-toxic and biodegradable. Can be used both inside and outside the house for many pests controls ants as well.

This cleaner controls insects, aphids, scale, mealybugs, ants and the whitefly (including the giant whitefly attacking San Diego) by means of suffocation, and or dehydration. Use of this product does not lead to the destruction of beneficial insects if used only on specific pest or plant.

Jungle Rain works very well against many types of funguses such as black sooty mold as well as for controlling ants outside. This mold is created as a direct result of insects such as aphids, scale, whiteflies and mealybugs. They suck the sugars from the plant, leaving a great deal of this nectar behind on the leaves of the plants being attacked. This sticky substance covers the leaves and provides nutrients for the mold to grow. This fungus interferes with photosynthesis, reducing the energy level of the plants being attacked. Ants too play an important role since they herd many sucking creatures around from plant to plant. Ants also have their own fungus that they carry on the bodies. This fungus will also grow on the nectar left behind by the sucking insects. Other insects will come around to feed on this nectar as well. **Jungle Rain's** two main ingredients are Castile Peppermint Soap and Citrus Oil. Smells great too! Many insects do not like this mixture of scents! Just follow instructions or formulas. Try 5 tablespoons per gallon water. Long Live Natural Soap!

USING DORMANT HORTICULTURAL OILS

Horticultural oils kill by suffocating insects and their eggs. Beneficial parasites and predators are not effected by this oil. This brand horticultural oil can be used on serious pests in the garden without upsetting the balance. Also helps to control many different types of fungus when sprayed during dormant period. The main problem is that they are not organic, made from pertro based and are therefore not recommend in the organic method. Vegetable oil works just as well.

Ultra Fine Oil: SunSpray is an Ultra-fine Horticultural spray oil. Again Not Organic!. A petroleum based product. This oil provides increased protection from phototoxity (the burning of plants leaves). This oil can also be used year round. Can be used up to day of harvest as well as year round! Also acts as a repellent keeping pests away for days after spraying. Ultra-fine oil is available from ARBICO, and through many mail order catalogs. Controls aphids, mites, beetle, leafminer, leafhopper, thrips, whiteflies, scale, psylla, tent caterpillar, borers, mealybug and more! Use as a last alternative! You will get same results from using vegetable cooking oils.

SPIDER STUFF

Spiders are a very important part of our eco-system. They will eat many times their weight in insects every day. While most true insects have six legs, spiders have eight. They will vary in size form very tiny to very large, depending on where in the world you happen to run across each other. I would rank spiders in fourth place behind ants (1st place), cockroaches (2nd place) and fleas (3rd place), but ahead of snails (5th) and gophers (6th) as the most feared and disliked pest. **There are more good beneficial spiders then harmful spiders.**

A good rule of thumb is that if the spider is not bothering you, do not bother the spider!

Here is a list of some spiders and how to control them on your plants and in the house:

Red Spiders: These tiny spiders are reddish in color and are found on indoor plants that have very little air circulation. Usually the plants are malnutrition and under watered. Overwatering will tend to damage the

WHITEFLY IN HOT WATER: Since 1991 horticulturist Jim Nichols Thousand Oaks, California in the United States has found that he can control whitefly with hot water. Nichols has filed a patent application for a line of equipment that uses a brief exposure to extreme heat and hot water for the control of whitefly, aphids, mealybug, scale and mite infestations.

Nichols has found a brief hot water treatment of 150-160 F. from one to three seconds controls all stages of the whitefly life cycle. For the five years that Nichols has conducted trials on his technique, it has been completely effective on whitefly and a broad spectrum of small bodied "infestation" insects without any damage to the treated plants.

Nichols is seeking interest and cooperation from the U.S. Department of Agriculture officials and from representatives of the agriculture industry for the most effective ways to expedite equipment development. For further information or comments contact:

Jim Nichols

2815 Hillman St.

Thousand Oaks, California, USA 91360.

plants more than the spiders would. To control, concentrate more on the health of the soil and the health of the plant and less on controlling the red spiders. However, if you have an unusually large amount of red spiders per plants, its time to make a few changes!

1...Repot the plant in a new compost, potting soil mixture. 2...Before you repot plants, bathe it in a mild soapy water solution. You can spray the solution directly on the spiders or you can bathe (place) the entire plant, pot and all into warm water with a dash of Dr. Bronners Peppermint soap. Harmless to humans.

Brown Recluse: Wear gloves when working in areas such as in garage, wooded areas, etc. Avoid contact whenever possible. Use soaps to control.

Arizona Brown Spiders: Avoid contact. Wear gloves when working near possible spider sites. Can be controlled by using soap such as Dr. Bronners Peppermint soap. Sprayed directly on spiders will kill them. Use before working in spider areas will prevent bites. DE dusted in areas will also control.

Black Widows: Should be left alone if possible. Males are harmless.

Tarantulas: Are harmless to human. Leave them alone is my best advise!

Letters to Andy

KITCHEN PESTS

Subject: Moths in the House

Question: Last year we had moths growing in a cereal box in the pantry and they literally took over. We wiped everything off with warm water, threw away a lot of grains, disassembled the shelves and painted the pantry walls. Initially the moths disappeared, but later reappeared, not in the pantry but all over the house. My spouse will not let me call an exterminator and wants to find a natural remedy. Do you have any suggestions? Thank you.

Answer: Description of Problem: There are many pests of stored foods, two of the most common are beetle and moth larvae. Also beneficials such as predators of these kitchen pests are known to inhabit stored foods as well. Cereal boxes are a favorite place for the Angoumois Grain Moth but also various types of grains, dog food, flour and almost anything edible. The fact that the problem went from locally (pantry) to all over the house suggests that you actually spread them by throwing away a lot of grains. It also suggests that you probably threw their eggs around and with any luck they made it through the winter to spring. Chemical controls are not healthy in the long term. Organic controls work better and last longer.

Causes: A particularly wet spring will encourage larger populations following spring. Also check infested stored food due to improper packaging and transport. Could be also at factory.

Immediate solutions: A clean up of all stored foods and the immediate area. Place new unopened packages into sealed containers if possible. Try storing in glass jars. Most people do not store their food stuffs properly inviting unwanted guests. A good rule of thumb is to go through all your stored foods at least once per year to avoid any problems. When you first buy your food items inspect for any opened packages, When you first open packages inspect for possible contamination. Use a natural soap and water solution to spray them if you disturb them from hiding. Use 5 table-spoons per quart water. DR. Bronners Peppermint soap is the best to use for this purpose but any natural soap will work.

Long term solutions: A light dusting of DE can be done around the shelves. You can place a liner on top. DE is not harmful if eaten so do not worry. The DE will last until next years spring cleaning! You can alterna-tively make a mixture of 1 part pyrethrum and 1 part DE and use that as a light dusting on shelves and places where thery are being seen. Take a close look around the house for any holes and other ways they can get into the house. A little repair here goes a long way.

Traps: Pheromone traps work well in catching them and giving you a good count of their population. A good sticky yellow trap will also work well. Another homemade trap is a bowl filled with vinegar. Just change regularly. That should do it!

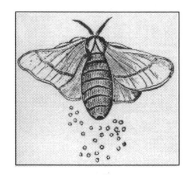

Moths
"A particularly wet spring will encourage larger populations."

WHITE FLY

Subject: white fly

Dear Andy:

This is my first look at your site and it's terrific! We are just getting into the agricultural business with a special growing soil found in China. Our new Agri friends say that the biggest problem here in California is the white fly. I am sure you are very familiar with this dangerous insect that destroys crop's en masse. Do you have any methods to curtail or eliminate this pesky fly. Appreciate your answer to this, or any referral to other sources. Regards.

Terra Firma Products, Inc.

Answer: Dear John, What is that special soil you are talking about? I would be interested in testing it for you and seeing for myself.... Con-cerning the whitefly problem. There are many organic solutions to this problem. My feeling is to deal with the cause and that the whitefly is only the effects.

Some common causes are:

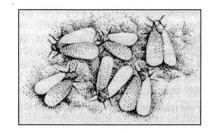

Whitefly
"DR Bronners Peppermint soap will kill them dead!"

◆ Overuse of chemicals both fertilizers and pesticides.. causes imbal-ance in nature....

◆ Overuse of high nitrogen fertilizers.. High nitrogen causes stress and rapid tender growth..

◆ City water kills bacteria in soil...dead soil causes imbalance amongst bacteria found in soil. Use a Water Transformer. See Appendix.

What is the answer?

1. Go Organic! Bring the soil back to life and nature will be able to control it herself, But that takes time which most people do not have...

2. Introduce natural whitefly predators such as Encarsia Formosa (I know it's for green houses but it does work outside too), a stronger predator is Eretmocerus nr. californicus a parasitic wasp.. Very effective! There is more info available, see my web site under Beneficials I have known.

3. Here are some organic tools you can use to kill them off with.......Dr. Bronners Peppermint soap will kill them dead! Experiment with the strength but 10 tablespoons per gallon should do it.. to this I would add a liquid pyrethrum/rotenone at 1 tablespoon per gallon (avoid if chemically sensitive since its is strong) and finally I would add a dash of Sunflower Oil (say 2 tablespoon per gallon). Avoid use during heat of day, best at night. Jungle Rain is perfect for this also. Use instead of Dr. Bronners if you have a bad problem.

4. You must control the ants since they herd aphids which produce nectar which the whiteflies love! See my web site entitled Dances with Ants.

Hope this helps.   Yours, Andrew Lopez

Concerning Wild Life

Deer, Rabbits and most wild life can be controlled by first understanding that they are living beings and have rights too. If we apply ourselves to allowing these creatures to live in harmony with us, we can provide for both us and them a better life and one in which we do not immediately think of how bad they are etc., but rather how fortunate we are to have them still here with us. I provide a feeding station for them on another part of the property that I do not mind them being in. Also there are various electronic devices such as Yard Guard (Arbico) that really work! Another helpful tool is your dog or cat since they can keep many animals at bay. A good product to use is called Deer Off. Athena Loucas (Deer Off inventor/housewife) says, "Everything in Deer Off is found in the kitchen, mostly food products. I'm not chemically minded. Chemicals work only temporarily." That's my kind of woman!

Deer Off works great on rabbits as well as most chewing animals and insects. Not a poison; all poisons should be avoided at all costs! Can also prevent dogs and cats from coming near your property.

---

19 You can always trust IGA members!

A NATURAL PEST CONTROL CHART by The Invisible Gardener

Note: Tobacco and Boric Acid are toxic if ingested; DE, Rock Dust, and Cayenne Pepper are harsh to the eyes and lungs; Soap, Garlic Oil, Hot Sauce or Dust may harm your eyes and lungs, avoid breathing or ingesting. Some people are allergic to Pyrethrum, Citrus Oils, Essential Oils. Any thing not listed please see index. For carpenter ants see ants.

Natural Pest Controls	Hand Pick	Traps	Ant Cafe	Snail Hotel	Stick-'ems	Jungle Rain	Dr. Bronner	N Soap*	Hot Water	Garlic Spray	Tobacco	Hot Sauce Spray	E Oil*	P*	Rock Dust	DE*	Hot Dusts	Herbs	R*	Cayenne	B*	Boric Acid
Ants			3			2	1	1	1	1	4	2	3	4	1	4		1				4
Aphids	1	2			3	2	1	1	1	1	4	2	3	4	1	4	3	1	4	4	3	4
Beetles	1	2			3	2	1	1		1	4	2	3	4	1	4	3	1	4	4	3	4
Borers	1	2					1	1		1	4	2	3	4	1	4	3	1			3	4
Caterpillars	1	2		3			1	1		1	4	2	3	4	1	4	3	1	4	4	3	4
Cockroaches		2	3		3	2	1	1			4	2	3	4	1	4	3	1		4		4
Fleas		2			3		1	1	1	1	4	2	3	4	1	4	3	1		4		4
Flies		2			3		1	1	1	1	4	2	3	4	1	4	3	1		4	3	4
Gopher/mole	1	2								1	4	2	3	4	1	4	3	1		4	3	4
Grasshopper	1	2		3			1			1	4	2	3	4	1	4	3	1	4	4	3	4
Grubs	1	2	3		3	2		1			4	2	3	4	1	4	3	1		4	3	4
J Beetle	1	2		3	3	2	1	1	1	1	4	2	3	4	1	4	3	1	4	4	3	4
Lawn Moths	1	2	3	3	3	2	1	1	1	1	4	2	3	4	1	4	3	1		4	3	4
Leafhoppers	1	2	3	3	3	2	1	1	1	1	4	2	3	4	1	4	3	1	4	4	3	4
Leaf Miners	1	2	3	3	3	2	1	1		1	4	2	3	4	1	4	3	1		4	3	4
Mealbugs	1	2	3	3	3	2	1	1		1	4	2	3	4	1	4	3	1		4	3	4
Nematodes		2	3	3	3	2				1	4	2	3	4	1	4	3	1	4	4	3	4
Rodents	1	2	3	3	3		1	1	1	1	4	2	3	4	1	4	3	1		4	3	4
Scales	1	2	3	3	3	2	1	1	1	1	4	2	3	4	1	4	3	1	4	4	3	4
Spiders	1	2	3	3	3	2	1	1		1	4	2	3	4	1	4	3	1	4	4	3	4
Spider Mites	1	2	3	3	3	2	1	1		1	4	2	3	4	1	4	3	1	4	4	3	4
Snails/Slugs	1	2	3	3	3	2	1	1	1	1	4	2	3	4	1	4	3	1		4	3	4
Thrips	1	2	3	3	3	2	1	1	1	1	4	2	3	4	1	4	3	1	4	4	3	4
Whiteflies	1	2	3	3	3	2	1	1	1	1	4	2	3	4	1	4	3	1	4	4	3	4

Column groupings: Traps and Handling (Hand Pick, Traps, Ant Cafe, Snail Hotel, Stick-'ems); Natural Sprays (Jungle Rain, Dr. Bronner, N Soap*, Hot Water, Garlic Spray, Tobacco, Hot Sauce Spray, E Oil*, P*); Natural Dusts (Rock Dust, DE*, Hot Dusts, Herbs, R*, Cayenne, B*, Boric Acid).

How to use this chart: #1 Do this first. Easy to make and safe to use. #2 Do this if you either did #1's and it did not work or you cannot do any of #1's. #3 Do this if #1 and #2 do not work. #4 Use as a last resort. Use with Caution, wear protective clothing, Avoid breathing.

N=Natural Soap, E=Essential Oils, P=Pyrethrum, DE=Diatomaceous Earth, R=Rotenone, B= Beneficials. Herb's are listed in this chapter and other parts of this book.

6

Organic Disease Control

If you Eliminate the Cause, You Eliminate the Effect.

When dealing with diseases, it's the cause that you must control, or cure, rather then the effect (disease). I will not attempt to describe each type of disease for you in this book. I will point you in the right direction since it has been my experience that you need not know the name of the disease to treat it. Remember, treat all diseases the same way. As long as you treat the cause, the effects will disappear. If the disease is still persistent after following the below steps, then proceed on to the organic disease control section.

ALWAYS KEY INTO NUTRITION FIRST.

What are you feeding your plants and soil? Tell me the truth now, really, what are you feeding it? If you say "nothing" then you are just as bad as the person who says "I am feeding it something I bought at the nursery"! There is nothing like proper nutrition for proper health! As in humans, so it is with plants.

Always Seek Balance. When things are out of balance, infestation and disease occur.

Balance is as important to the soil and plants as it is to humans. When things are in balance, all form of stress is reduced accordingly.

The Good Guys and the Bad Guys do not live in the same place.

Bad bacteria and good bacteria do not share the same soil. Dead soil is not actually dead but instead contains the presence of bad bacteria or fungi or any number of disease sources.

AVOID HIGH NITROGEN FERTILIZERS.

The continual use of chemicals in the soil eventually destroys both the bacteriological and biological balances of the soil. Also present are the salt deposits of the various fertilizers, etc., used through out the years. This causes major changes in the PH levels of the soil effecting nutritional absorption by the plants. This will lead to higher stress levels in all things associated with this soil. The application of not only high nitrogen fertilizers but the application of any type of fertilizer is very bad for the soil, the plants, the insects, the animals, the birds and all humans involved. High nitrogen also promotes the various diseases found in the garden. Why? High nitrogen causes stress & imbalance in the soil & plants, the same way a high sugar diet does in humans. The more natural the sources, the more acceptable the supply.

Poor health and high stress of plants are due to improper nutrition.

The stronger the plant the less the disease will effect it. Avoid using any chemical fertilizers, pesticides, herbicides, etc. Chemical fertilizers are not complete foods. They provide only a small amount of the total diet required for healthy plant growth. NPK does not

Diseases in different plants may have the same name but will show different symptoms thus making identifying very difficult if not impossible. While I encourage you to try to identify the disease that is attacking your plant(s), you must remember that the basic treatment(s) remains the same regardless of the disease.

"Fish Emulsion: Fish emulsion will provide essential oils that will prevent many diseases from occurring."

a complete dinner for plants make!

STRESSED OUT PLANTS.

The same rules of pest control apply to disease controls, the higher the stress, the greater the problem associated with it and the harder it will be to control.

DEAD OR WATER SOAKED SOIL.

Amend with compost, rock dust and animal manure if available. You should have plenty of earthworms. Dead water soaked soils have earthworms. Compost and mulching will increase organic matter to soil allowing bacterial and nutritional resources to increase. Increase drainage. Practice soil rotation if possible. Rotation of crops is best whenever possible. If a raised bed, then allow the bed to go fallow for one season.

SOME DISEASES INTRODUCED BY VIRUSES.

Destroy diseased plants. Throw away in the trash. Do not compost since most backyard gardeners do not compost in big enough piles in which the temperature is high enough to kill the fungi. Avoid spreading the disease by cleaning your tools with alcohol inbetween cuttings. Do not use chemical fungicides. They only will cause more problems both to your health and the soils health[1]. Learn to use Compost Tea.

SOME DISEASES INTRODUCED BY WEEDS.

Use disease free seedlings and plants. Diseases carried from one plant to another by humans, insects, animals, etc.. Be specially careful when bring potted plants with soil into the home. Certain weeds are hosts to pests that carry the disease from one plant to another, therefore weed control of some type is important. Weeds can also provide homes for many beneficials so choose carefully.

SOME DISEASES SPREAD BY INSECTS.

Many diseases carried around by insects such as the ant. Control the insect populations in your areas. Learn what insects spread what disease.

IMPROPER VARIETIES PLANTED.

Proper placement of variety is important. Choose a location that gets as much light as needed by the plant for proper growth. Make sure the plant(s) will grow normally in your area. Choose the correct variety for your area. Heirloom varieties are stronger and more suited for organic gardening. Write to Abundant Life Foundation or Seeds of Change. See Resources Directory for their address. Use Trap crops in addition to various types of traps that attracts the pests away from the garden.

IMPROPER PLANTING TIME.

Observe the correct planting dates on vegetables, etc. For best results. Planting during the wrong time of year also exposes plants to insects that would not normally affect your plants, if they planted at the right time. Also temp and amount of day light are important.

---

1 Chemical Fungicides destroy the bacteria & "Goods Guys" in the soil, and can cause health problems in humans and animals.

IMPROPER WATERING TECHNIQUES.

Avoid overhead watering. Use a soaker hose or a drip system whenever possible.

How Important is it to Identify the Disease?

Diseases in different plants may have the same name but will show different symptoms thus making identifying very difficult if not impossible. While I encourage you to try to identify the disease that is attacking your plant(s), you must remember that the basic treatment(s) remains the same regardless of the disease. This book was not designed to help you to identify the pest or disease. There are many books out there that are good for that purpose.

NATURAL DISEASE CONTROLS

Minerals: There are over 72 minerals that are needed by the soil and the plants for healthy disease free growth.

Bordeaux: Is a mixture of copper sulfate and lime. Used in France for at least 100 years. Very poisonous, be careful applying. Follow instructions on label. Good for fungus diseases on grapes, fruits such as peaches, apples. Used against Potato blight. Helps' roses with fungus. Available from ARBICO, etc. And at most nurseries.

Calcium: Calcium is a natural fungicide. Applied either directly on the leaf or into the soil. Calcium reduces certain types of bad bacteria, fungi found in many diseases. Also beneficial bacteria enjoy the calcium in their diet. So a good spray will naturally be a mixture of calcium and a bacterial source. Calcium obtained from crushed egg shells. Bone meal is also a good source of calcium. Milk provides calcium and a beneficial bacterium (see Milk). Rock Dust is high in calcium (see Rock Dust and Kelzyme).

Copper: Copper is an excellent natural fungicide, controls diseases of vine crops, potatoes, leaf spot, anthracnose, downy mildew, powdery mildew, scab, fire blight, bacterial spot. Available in liquid or dust form. Available from ARBICO, etc. And most nurseries. Use sparingly. Prevent black spot on roses, tomatoes; and other diseases. A rotenone and or copper mixture will also control insects. Excellent for use on melons and cucumbers. Can be used as a liquid for spraying or as a dust for dusting.

Magnesium: Is available in Epsom salts. Available at most drug stores. Magnesium will help to reduce stress on the plants. Plants can become stressed out due to minerals not being available to them. High nitrogen causes imbalances in the soil, and plants' bio-mechanisms, resulting in the locking up of trace minerals in the soil, and the inability of the plant to intake the required minerals. You can apply at the rate of ½ cup Epsom salts to 1 gallon water, sprayed on leaves.

Potassium: A great stress reliever for plants. Increase absorption of calcium and magnesium. A natural fungicide.

Sulfur: Is an ancient tool of the organic farmer. A natural fungicide and insecticide. Used for rust, powdery mildew, leaf spot, brown rot. Also a good soil acidifier. Available from ARBICO, most nurseries. Micronized

The Invisible Gardener says:

"It does not take magic to control diseases naturally!"

sulfur is an extremely fine dust which when dusted on plants will control apple scab, cedar-apple rust, black rot, leaf spot, and powdery mildew. Liquid sulfur is also available.

Zinc: This exotic element is just as important as the top three (NPK) without which the plant would die. Zinc is normally readily available in the soil but when the flow minerals needed by the soil and plant is interrupted, the zinc will be used up by the plants and quickly run are depleted. When the plants do not have any zinc left they go into immediate stress. Good rich compost will provide the zinc.

Natural Sprays

Alfalfa Tea: A tea can be made using Organic Alfalfa meal. I suggest that you add 1 cup per gallon of the meal. Allow to sit for 1 hour, strain thru a gardener's cheesecloth. Spray on plants to control most fungus before they start. Add a dash of natural soap as a wetting agent. Great as a blend with any of the minerals mentioned above. Alfalfa is available from Arbico, Nitron, etc.

Baking Soda: Baking soda made into a spray at the rate of 1-5 tablespoons per quart (depending on the plant sprayed). Baking soda and milk with a dash of natural soap makes an excellent fungal control. See chart for list of diseases it controls. A better mixture is in a quart of compost tea water add: 1-5 tablespoons baking soda, a dash of natural soap, a dash of any natural vegetable oil, 1/2 cup of vinegar or apple cider.

BD 508: A biodynamic spray that will help to prevent many fungal diseases. See Disease Chart. This spray is available from the Bio-Dynamic foundation. See resource directory for address.

Clay: Clay can be used as a fungal control. Add 1 cup of finely powdered clay into a gallon of filtered water, allow to sit for an hour. Strain thru gardener's cheesecloth. Add seaweed to increase effectiveness. Use either red or white clay.

Compost Tea: Compost tea when properly made will control many fungal diseases. Place 1 cup of your favorite compost into a panty hose. Tie into a ball. Allow to sit in filtered, solarized, spring or rain water. The length of time depends on the severity of the problem. Add 1/2 cup Molasses. Allow to sit for approx. 1-5 days. Add a dash of natural soap. Then spray on plants, controls ants and most insects as well as various diseases in addition it fertilizes!. See index for more info.

Fish Emulsion: Fish emulsion will provide essential oils that will prevent many diseases from occurring. Use 5 tablespoons per gallon. Avoid Urea based products. High nitrogen is not good for disease control. Good sources of fish emulsion are Nitron, or Arbico or Gardeners Supply.

Garlic: Has been used for centuries as a fungal control. Garlic can be made into an extract and sprayed. Add a dash of natural soap to increase effectiveness. Garlic can be grown around the plants. The natural aroma will provide for a fungus free environment. This is mainly due to the garlic being absorbed into the plant.

Horsetail: An old time favorite of Bio-Dynamic gardeners. Affective against many bacterial diseases. An infusion of the leaves, flowers is O.K.

"The soil is like a bank account. You must put in and put in and wait for the interest to grow before you start making withdrawals. Who has a bank account anywhere where you can only withdraw and never deposit?"

Don Eligio Panti

For most problems; for greater strength make an extract or obtain thru mail order companies. The dried herb is available thru many mail order herb companies. For better results mix with a dash of soap and a dash of garlic.

Hydrogen Peroxide: This product will reduce many types of fungal diseases. Use only as a temporary control. Use 1-5 tablespoons per gallon water of the 3% (best to use food grade at 1 drop per gallon). Always test for strength so as not to burn plants. Always use filtered water when making any natural bacterial sprays. Use seaweed or natural soap to increase effectiveness.

Lime: Lime will control many diseases around the garden. Use with care as it will kill off beneficials as well. Make as a spray.

Milk: The high calcium levels combined with milk's natural bacteria make for an excellent natural spray to control fungal diseases on many plants. Add a dash of garlic, a dash of baking soda and a dash of soap. Raw milk is best but otherwise it does not make a difference the amount of fat it has. Powdered milk will work if it's only the calcium you want. Use 1 cup per quart water.

Manure Tea: Manure tea makes an excellent antifungal spray as it introduces many different types of beneficial bacteria. Avoid using raw manure's. Use only well aged. See manure tea in appendix for preparation. Add a dash of natural soap to increase effectiveness. Manure placed into panty hose, tied into a ball and allowed to sit in filtered water.

Molasses: Will instantly raise the energy level of the plant. This instant raise will help to deter many fungal diseases and pests. Keeping pests away is a very important part of controlling diseases naturally. Add a dash of either natural soap or fish emulsion to increase effectiveness. Go well with milk and also with rock dust. A good strong anti fungal formula is equal amounts of molasses, seaweed powder, milk and rock dust (about ¼ cup ea.), Mix well until a paste. You can then place small amount of the paste into a panty hose and tie into a ball. Place ball into a gallon water. Allow to sit for 2-4 hr. Strain water then spray on diseased plants. A good store bought molasses is Granny Smith's molasses.

Rock Dust Milk: Rock dust has very large amounts of calcium, iron, magnesium as well many minerals that make it a quick energy boost to plants and increasing the beneficial bacteria count in the soil. Rock dust added to filtered water and turned into a milky liquid and sprayed or used as dust. I suggest adding a dash of natural soap along with a dash of seaweed. A dash would be 1 tablespoon of powdered seaweed or 1 ounce of liquid seaweed per gallon of filtered water. See rock dust in appendix for its preparation. Best to let the rock dust settle down to the bottom of the water then strain water to avoid clogging sprayer.

Vegetable Oil: Can be used as a way of controlling many diseases when they first occur. Should be used during warm or cool weather. Avoid using during hot weather. Corn oil, soy bean oil, olive oil, coconut oil are but a few oils you can use.

The Sun is one of the best natural controls for many diseases since they cannot withstand the sunlight!

Milk: The high calcium levels combined with milk's natural bacteria make for an excellent natural spray to control fungal diseases on many plants.

Natural Disease Control Chart *by* The Invisible Gardener

1. Try Making it yourself first! See Index for formulas through out book.

2. Do this second if you cannot make #1. Local Nursery or Store or Hardware.

3. Something's you cannot buy at your local store so order from catalogs. All listed are IGA members. See resources for address for ordering.

4. Use as a last resort. Handle with caution and proper protection as they may make you sick or can injure beneficials, etc.

Natural Disease Controls	Compost Tea	Milk	Copper	Liquid Sulfur	Liquid Garlic Spray	Lime Sulfur	E Oil*	Vegie Oil	Manure Tea	Liquid Seaweed	Nitron A-35	Agri Gro	Acadie Seaweed	Rock Dust Milk	Molasses	Baking Soda	H2O3	BD 508	Cal- cium Spray	B*	H*	Super Seaweed
Anthrac	1	1	4	4	1	4	2	2	1	2	3	3	3	1	2	2	2	1	1	3	1	3
Apple Scab	1	1	4	4	1	4	2	2	1	2	3	3	3	1	2	2	2	1	1	3	1	3
Bean Rust	1	1	4	4	1	4	2	2	1	2	3	3	3	1	2	2	2	1	1	3	1	3
Gail Formers	1	1	4	NA	1	NA	2	2	1	2	3	3	3	1	2	2	2	1	1	NA	1	3
Bacterial Blight	1	1	4	4	1	4	2	2	1	2	3	3	3	1	2	2	2	1	1	3	1	3
Bacterial Canker	1	1	4	4	1	4	2	2	1	2	3	3	3	1	2	2	2	1	1	3	1	3
Bacterial Spot	1	1	4	4	1	4	2	2	1	2	3	3	3	1	2	2	2	1	1	3	1	3
Blossom Drop	1	1	4	NA	1	NA	2	2	1	2	3	3	3	1	2	2	2	1	1	NA	1	3
Black Rot	1	1	4	4	1	4	2	2	1	2	3	3	3	1	2	2	2	1	1	3	1	3
Common Mosaic	1	1	4	4	1	4	2	2	1	2	3	3	3	1	2	2	2	1	1	3	1	3
Damping Off	1	1	4	4	1	4	2	2	1	2	3	3	3	1	2	2	2	1	1	3	1	3
Downy Mildew	1	1	4	4	1	4	2	2	1	2	3	3	3	1	2	2	2	1	1	3	1	3
Fire Blight	1	1	4	4	1	4	2	2	1	2	3	3	3	1	2	2	2	1	1	3	1	3
Early Blight	1	1	4	4	1	4	2	2	1	2	3	3	3	1	2	2	2	1	1	3	1	3
Fusarium Wilt	1	1	4	4	1	4	2	2	1	2	3	3	3	1	2	2	2	1	1	3	1	3
Powdery Mildew	1	1	4	4	1	4	2	2	1	2	3	3	3	1	2	2	2	1	1	3	1	3
Orange Rust	1	1	4	4	1	4	2	2	1	2	3	3	3	1	2	2	2	1	1	3	1	3
Rhizoctonia	1	1	4	NA	1	4	2	2	1	2	3	3	3	1	2	2	2	1	1	3	1	3
Scab	1	1	4	4	1	4	2	2	1	2	3	3	3	1	2	2	2	1	1	3	1	3
Sunscald	1	1	4	NA	1	NA	2	2	1	2	3	3	3	1	2	2	2	1	1	3	1	3
Tobacco Mosaic	1	1	4	4	1	4	2	2	1	2	3	3	3	1	2	2	2	1	1	3	1	3
Verticillium Wilt	1	1	4	4	1	4	2	2	1	2	3	3	3	1	2	2	2	1	1	3	1	3
Leaf Spot	1	1	4	4	1	4	2	2	1	2	3	3	3	1	2	2	2	1	1	3	1	3

E is Essential Oils, B is for Bordeaux, H is for Herb's , All #1's can be mixed together with out any damage, all others follow formulas.. To use this chart first find disease then locate what control is #1 that you can use then locate that control in index that will refer you to the page where the formula is. You can also look up the disease in index first which will point to this page plus page of formulas for that disease. Get it? Boy I need a rest!

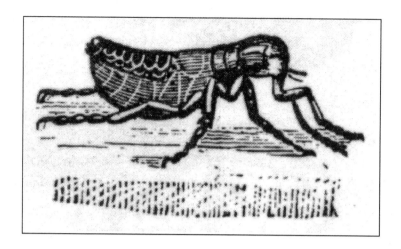

7

Natural Flea Control

About Fleas

Let's talk about the cycles of the flea. There are three different types of fleas, the Cat Flea (Ctenocephalides), The Dog Flea (Ctenocephalides) and The Human Flea (Pulex irritans). The Cat Flea is the one we are most interested as the other two are very rare. Cat Fleas not only like cats, dogs, and humans but they also like many other species such as rats, chickens, etc.

THE LIFE CYCLE OF THE CAT FLEA

Adult fleas lay eggs outside in the grass or any warm safe place (usually near a warm blooded host). The eggs develop into legless larva that feeds on tiny amounts of organic matter for up to one month. Usually the eggs are laid on the hosts that then drop to the ground. If inside they will fall on the carpet, etc. The larvae do not bite but live off dried blood defecated by the adult fleas. Outdoors the larva will lie in a shady moist spot. The larva then becomes a pupa by spinning a cocoon and emerges in about a week as a hungry adult. The cycle for reproduction is 30 to 60 days and is triggered by the presence of the host's warmth and the carbon dioxide given off by the host's respiration. The Larvae can live up to 6 months and the pupae can live to a year until the conditions are right for emergence. What this means is that if you have just moved into a new apartment or home which previously had a dog or cat as a tenant, that your presence, or your cat's or dog's, could trigger off their emer-

gence! Many people have allergic reactions to flea bites while others never notice it. Fleas usually do not bite humans unless there are no other mammals around.

CONTROL METHODS

Your treatment should be spread out over a 60 day period. This means that you should do regular flea treatments every month if you are in areas prone to having fleas. I have two cats and one dog. I know fleas. The cats clean themselves and generally do not like to be told what to do or to have anything done to them, but they can be helped as well as the dogs. Fleas and dogs, fleas and cats, they go together. It's unnatural for there never to be fleas again. WE cannot totally rid ourselves of them any more than we can totally rid ourselves of the ants. The answer here is to control and reduce them. The formulas given here along with the hints are to help you keep fleas out of the house and off your animals. Follow these hints and you will succeed. Many flea collars on the market contain Sevin that attacks the nervous system of the insect (as well as your pet's nervous system and ours as well). I have seen flea collars which contain DDT! The companies are getting smart and are using Pyrethrum[1] but are screwing up when they add Piperonyl Butoxide to boost it up. This product has been asso-

---

1 For more info on pyrethrum see index

ciated with liver disorders. You should avoid using this product. Especially avoid using around children or seniors or those with health problems. Insist on using only pure pyrethrum flowers (the whole plant). Read the ingredients. ARBICO, REAL GOODS, GARDENER'S SUPPLY are a few places you can buy pure Pyrethrum (see resources directory). Also available at many garden centers, etc.

Steps to Organic Flea Control

Step 1: Spray'em

The first thing you have to do is to reduce their population ASAP. This is done by spraying the infested areas with a natural soap that will kill the fleas (but not the eggs). This can be sprayed around pet bedding, rugs, furniture, etc.. A fine misting is enough. Should be done daily. Always test rugs, furniture for any damage the soap may cause.

USING NATURAL SOAP SPRAYS

Dr. Bronner's Peppermint Soap: This is an excellent soap to use as a mist around infested areas. Smells great, too! See Index for the various places this soap is mentioned in this book. See also soap page in index. Use 5 tablespoons per quart sprayer of water.

Safer Flea and Tick Spray: Pyrethrins and insecticidal soap are combined to make an effective spray. Add the Dr. Bronners or Green Ban to make an even more powerful mixture. Add equal amounts of each or you can experiment for strengths. Spray around infested areas. Safer also makes an Insecticidal Flea soap for dogs to bathe in.

Step 2: Dust'em

USING NATURAL DUSTS

There are several different dusting formulas which you can use on fleas, this will depend on your preference. You dust lightly before and after you vacuum.

A Good Natural Flea Control Formula with Pyrethrum: Pure Pyrethrum powder is one of the safest ways to kill and control fleas both on and off animals as it is totally harmless to humans and animals. Avoid breathing dust or fumes. Comes from the Pyrethrum plant. Dust dogs once or twice per week, for a bad case use daily. Dust dog bedding and rugs, carpets etc., before vacuuming. A good approach is to start in one corner of the room and dust slowly as you work your way out. Allow to settle for 1 hr before vacuuming. Real Goods and Arbico both sell a great pyrethrum powder.

DE and Salt Formula for Fleas

1 lb. Dia-Earth

10 oz. Salt

Inside: Mix Dia-earth with the salt and lightly dust the rugs. You can use a salt shaker to dust with or make your own. Work into rugs with a brush. Avoid breathing. Allow to sit for a 1 hour then vacuum. You can

"Dr. Bronner's Peppermint Soap: This is an excellent soap to use as a mist around infested areas. Smells great, too! See Index for the various places this soap is mentioned in this book. See also soap page in index. Use 5 tablespoons per quart sprayer of water."

lightly dust afterwards and leave it on. This will prevent further infestation. Dogs can be dusted with pure Dia-earth. Dust once per week as needed. Use a small handful of DE only and rub on coat. Be careful to avoid the eyes.

Outside: Lightly dust outside areas once per month, using only the DE as the salt would not be good outside. A better outside control method is nematodes. See Bio-halt in index.

Dia-Earth Pyrethrum Formula

10 lbs Dia-Earth

1 lb. Pyrethrum Dust

Mix the Dia-earth with the pyrethrum powder. This is a very strong combination and will kill and control many other insects as well as fleas. Can be used inside or outside. Inside use very lightly around pet bedding, sleeping areas, frequently used areas, etc.. Outside use only in areas where you know there are fleas. Use only sparingly. Diatomaceous Earth is available at: ARBICO, and NITRON INDUSTRIES, GARDENER'S SUPPLY, PEACEFUL VALLEY, and many mail order companies.

Using Dried Pennyroyal or Peppermint: Sprinkle dried Pennyroyal or peppermint on rugs, animal bedding, etc. Can also be made into a tea and used as a bath for dogs. Grow more herbs in your environment. Growing herbs such as Pennyroyal, peppermint, etc. will naturally repel fleas, etc. Pennyroyal is a fragrant low growing ground cover. Easy to grow and effective against fleas. You can alternate between any of the above mentioned dusts. Do not use anything that does not feel right for you. Always test a small portion of the rug to avoid color damage.

Step 3: *Vacuum, Vacuum, Vacuum*

The treatment is regular vacuuming of rugs, carpets and upholstery, pillows, mattresses. etc. Be sure to vacuum the pets' sleeping areas as well. Vacuum at least twice per week, daily is better for bad flea cases. Be sure to properly dispose of the vacuum bag otherwise the fleas will get out (tape closed). Vacuuming helps not only to rid yourself of the adult fleas but the eggs and larvae as well. I realize that this is a lot of work but it's necessary for proper control of fleas. When fleas are reduced then you can vacuum once or twice per week or as often as necessary. I am not saying that the answer to fleas is vacuuming, though it does help.

Step 4: *Using a Flea Comb*

There are special metal flea combs (better then plastic type) available that you can use to comb your cat or dog. This is a very effective method of controlling and reducing fleas. Combs monitor fleas control the problem by picking up flea eggs. You dip the flea comb into soapy water or you can add a little alcohol to the water, to kill the fleas. Comb during bad flea periods. Comb once or twice per day if you can. Best for cats. Feeding a little garlic to your pet will help them rid themselves of fleas, and it will also control worms. Feeding brewer's yeast to your animals will also help control fleas on them. An herbal flea collar would be

" The first thing you have to do is to reduce their population ASAP. This is done by spraying the infested areas with a natural soap that will kill the fleas (but not the eggs)."

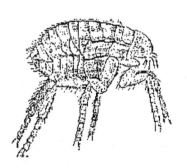

Green Ban Dog Shampoo

"*One of the best controls for fleas is proper nutrition and exercise*"

---

very useful here. Sometimes, electronic collars work well.

Step 5: Safe Shampoo for Dogs

Bathing your dog regularly with a natural based soap will go a long way towards reducing pests on your animals. Always read the ingredients before buying. Avoid using any soaps with additives or coloring, dyes and in general avoid using if you cannot pronounce the ingredients! Be careful using shampoos that use pyrethrum with Piperonyl Butoxide. It's not the Pyrethrum that you should avoid but the additive. This stuff (PB) is harmful to your dog and you as well. A good safe shampoo is called Flea Stop that contains D-Limonene (comes from lemon peels) which kills fleas dead.

Dr. Bronner's Peppermint Soap: Bathing your dog in Dr. Bronners Peppermint soap will kill all fleas and ticks while making your dog smell nice. You can also add a teaspoon of Dr. Bronners into a quart sprayer filled with water and use to spray or mist your dog before he comes into the house. Use a towel to dry This mixture can also be sprayed around the dog's bedding area and around the entrances to the house as mentioned in step 1.

Green Ban Dog Shampoo: This shampoo comes from down under. This soothing, fresh smelling conditioning shampoo contains the pure natural essential oils of peppermint, myrrh, eucalyptus and cajuput, plus apricot kernel oil and extract of kelp in a base of non-detergent coconut shampoo. This product contains all the right ingredients that make for an effective natural spray (sounds like an ad does it not?).

Tea Any One?: Next time you make peppermint tea, save the tea bag. Next time you bathe your dog, make a batch of peppermint tea. Allow to cool and use as a rinse on your dog.

Step 6: Birth Control

Continual use of the dusts, sprays and vacuuming mentioned above will provide a form of birth control for fleas. I suggest dusting once per month during flea season and spraying as often as needed. Also, see Nematodes. There are several new products on the market for flea control that works by applying a liquid on their backs and it is absorbed into their system. The word is still out (for me) how safe they are. Any one has info on this one way or the other please let me know.

Bio Flea-Halt: A relatively new nematode product that controls fleas outside! Using specific beneficial nematodes to control flea larvae and pupae. It will effectively kill over 90% of both the larvae and pupae within 24 hr. Safe to use around the vegetable garden, lawns, etc. Contact Farnam Companies of Phoenix, AZ for more info on local sources.

Step 7: Here Fleas, Here Fleas... The Ultimate Flea Trap

Fleas like most insects are attracted to light. A trap can be bought that uses a light bulb and a sticky strip as the base. This trap works at night in a dark room. The fleas in the carpets are attracted to the light and are stuck on the sticky mat. There are many companies now selling this prod-

THE ULTIMATE FLEA TRAP

uct. A simple homemade version would be using a light clamp, shining the light onto a sticky mat. You can also use a bowl of soapy water and shine the light on the water or you can use cheap wine. The wine will also attract them. Shining light on the wine will make the wine scent travel further.

SOME HELPFUL HINTS

For long term control of fleas on your property, you must strive to constantly raise the energy level of your property. This is done by providing your soil with plenty of good rich compost, by regular mulching and by promoting a biological diversity. It will be this diversity that will automatically control the infestation of any one insect.

Stress can make an animal or human more prone to flea attacks. Reducing stress will reduce pest attacks. The principle laws of balance are the same in all living things. From the smallest to the largest, all living things require balanced energy. The equation of life dictates that harmony be balanced by evolution or growth. In animals, this balance is maintained through proper nutrition, proper elimination and proper exercise. Keep yourself healthy by eating and exercising. Keep your pets in shape by feeding them good food and giving them plenty of exercise. Read the ingredients of the food you are feeding your animals. Avoid foods with chemicals as additives.

MORE SPECIFIC ORGANIC FLEA CONTROLS

Natural Insecticides for flea control are available as liquid solutions, forgers, powders, and aerosols. It is important that you understand what their main ingredients are and if any harmful additives are present.

Here is a list of safe natural products used in flea control:

Pyrethrum: Pyrethrum is made from the crushed pyrethrum chrysanthemum flowers, and is safe to use around mammals (kills fish), safe for birds' pets, etc. Pure pyrethrum is the best to buy (avoid additives such as PB. See below). Arbico and Gardeners Supply both sell a good pure pyrethrum product. Dust only your dog(s) with the pyrethrum. Cats lick themselves and would not like its taste. Avoid eyes. See index for more info.

Pyrethrins: Pyrethrins is the active insecticidal compound found in Pyrethrum. Pyrethrum comes from a flower and is safe to mammals. Many products are available for flea control containing pyrethrins, however be warned that most products also add the synergist Piperonyl Butoxide, which has been found to cause Liver disorders and can cause chronic human health problems. Children, elderly and pregnant women should avoid completely. Safer Entire Flea and Tick spray does not contain this PB. Always check the label to see if PB is present. Please see index for more info.

Repel: There are many natural repellents available on today's market. Some of the best are made from essential oils.

Here are a few Essential Oils you can use: Pennyroyal Oil, Peppermint Oil, Eucalyptus Oil, Citrus Oil, Citronella Oil, Rosemary Oil, Tea Tree Oil. These oils can be used in conjunction with flea collars or by spraying a mild solution (1 drop per quart water) on bedding, etc. I do

Stress can make an animal or human more prone to flea attacks. Reducing stress will reduce pest attacks. The principle laws of balance are the same in all living things. From the smallest to the largest, all living things require balanced energy.

"For long term control of fleas on your property, you must strive to constantly raise the energy level of your property. This is done by providing your soil with plenty of good rich compost, by regular mulching and by promoting a biological diversity."

not suggest putting on dog's coat. Instead you can add one drop on a cloth and use the cloth to wipe the dog's coat. Wipe again with a clean cloth. Cedar chips make a good flea repellent when placed around the animal's bedding. Replace with new cedar chips regularly as with age it loses its odor and may become a good flea-larvae home.

Limonene/Linalool: A citrus-peel extract, D-Limonene and Linalool are both safe to use on mammals. Products containing only d-limonene kill larvae and adult fleas, while products that contain both kill all stages of the flea. I suggest that you use lightly as some pets may show skin rashes.

Using Boric Acid: Boric acid can be used in the home to control fleas. It can be safe if you follow instructions. Just remember that if you have pets that it can kill them as well as kill animals including humans if ingested or if it enters our blood. It is for this reason that I urge great caution when using this product. Borax laundry soap works great for this purpose because it has a fragrance with it that prevents animals from eating it. It nevertheless has to be used with caution and you must avoid breathing. Basically you apply a small amount of the boric acid onto the rug and then you brush into rug removing any left over dust with a broom. This dusting will kill any flea eggs as they hatch and will last for a long time. You can add salt to this mixture at 50/50 to increase effectiveness. See index for more info on boric acid. See also page 13.

8

Organic Cockroach Control

About Cockroaches

Second only to ants in instilling hate and fear in all living things, cockroaches have been called " The Rats of the insect world". Cockroaches belong to the order Blattoidea (family Blattidae), and can live up to a year. Females lay several egg cases containing 30-50 eggs ea. during her life time. Some varieties carry their egg cases with them and place them for hatching later. Cockroaches utilize their antennae to pick up chemical signals from the air. This helps them in keeping away from any synthetic chemical used against them. What this means is that cockroaches will detect pesticides and avoid the area. They also sample food stuffs before eating to detect pesticides used against them. This is why using chemicals against them is fruitless. Cockroaches have as good a memory as ants. Like the ant, they too have developed immunity to many pesticides.

Since the first cave people, roaches have shared their home with humans. Cockroaches will live in almost any environment. They can be found on board ship, near food areas, bathrooms. Their eggs are good travelers, hitch hiking on board container's en route. Air travel does not bother them. They will thrive any where the moisture, temperature and amount of food available falls within tolerable boundaries. A cockroach is capable of traveling many miles in one day. They can move through the cracks in walls. Where they go, their eggs will follow. A cockroach can live without food for three months and without water for over thirty days. Their egg cases can survive for years under the right conditions. There are over 4,000 species of cockroaches known, 57 are found in the United States. Cockroaches have existed since the Paleozoic era that is about 400 million years ago. Only 6 species in the USA are considered bad household pests.

Steps to Organic Cockroach Control

Step 1: Using a Natural Cockroach Dust

20 Mule Team Borax: This is not a soap but an additive (laundry booster) made from borax. It is a great natural product to use for the laundry but this product can be used with great success in controlling roaches. Simply sprinkle the borax powder in places where roaches are hiding (attic, insides' walls, drawers, cabinet, car port, etc. Places where your kids, dogs, cats, etc. cannot get at). The main caution is to be careful where you put it so that it is not ingested by anything other then the roaches. Use lightly. Also avoid breathing or getting on cuts. See boric acid in index.

A Safe Dust Formula

1 cup peppermint leaves (powdered)

1 tablespoon DE (garden grade)[1]

1 tablespoon of cayenne pepper or chili pepper,

1 tablespoon powdered pyrethrum

1 tablespoon salt

---

[1]DE can be used by itself as a dust for roach control. Use as little as possible.

Coconut Oil will coat the cockroaches and will suffocate them. Add a dash soap to this and increase its effectiveness. Experiment for proper strength.

Citronella is an incredible tool in cockroach control. Learn to use this oil against cockroaches. Using a small amount in water will repel them. Garlic oil can be used against cockroaches with excellent results, (If you do not mind the garlic scent).

Add a few drops to soap and water. Spray directly on cockroaches.

Melaleuca oil is a very strong oil to use against cockroaches. Use a few drops with a dash of soap and water. Test for strength.

Peppermint oil is another strong oil. This will kill cockroaches on contact! Test for strength and scent tolerance.

Blend well together. Use a mortar to grind together into a fine powder. Sprinkle in locations cockroaches are seen. Spread a very thin layer. Follow along walls, in cracks, under structures. Be careful using cayenne pepper as it will make you sneeze. Use is optional (works best along outside or unused areas). Use as much as you can handle. African Cayenne pepper is 940 BTU's[2] and the best to use. Any type of pepper will do. DE should be used lightly inside. Works best if allowed to stay for 24 hr. or more. In between walls, under houses, use in the attic, and in the cellar, are good places to apply. Put in closet corners. Use outside along side of house and dust in places where they hide (under logs, wood, etc.). This mixture can be made into a paste by adding a little bit of water, and a tablespoon of butter, stirring until a paste like mixture is made. As a paste, you can place into roach cafes and place in the areas they are seen (but hidden). They will eat this and die.

Step 2: Using Soaps[3]

Dr. Bronners Peppermint Soap: A safe natural soap. In use for over 30 years as a bath soap. Very interesting label, fun to read! Dr. Bronners makes several different types, peppermint, lavender, eucalyptus and others. The scent of these soaps makes them very effective against any insect. Dr. Bronners Peppermint soap is a very effective tool against roaches. It is an environmentally safe soap to use. Spraying directly on roaches will kill them. Experiment on strength to use. Try 1 capful per quart. Use in kitchen and where roaches are seen. Spray along their paths to repel. Use 1 tablespoon per quart water. They can not develop an immunity to this. Many people prefer this because it has no scent. A concentrate, so use a few drops per quart. Sprayed directly on roaches. Citrus Soap is a safe natural soap to kill cockroaches. Experiment with strength needed to kill them. Try a capful first. Herbal Soaps can be a very important tool in your fight against the cockroach. Read the label first then try it. Insecticidal soaps made from fatty acids and are excellent for soft bodied insects. Safer Insecticidal soaps can be blended with any of the above soaps to increase effectiveness.

Tabasco Soap: Tabasco sauce and soap work great as a liquid repellent sprayed outside. Simply add 2 tablespoons Tabasco sauce into a quart spray bottle filled with water and a tablespoon of a natural soap (such as Dr. Bronners Peppermint soap). Spray along where the walls of your house meet the ground. Spray also in areas outside where they hang out. This will keep them away. This mixture can also be sprayed directly on them to kill them.

Step 3: Roach Traps

Sticky box traps: A sticky substance is used inside an open end box. They are called Roach Motels and can be purchased at most garden centers. Pop bottles make excellent traps as the fermentation of the soda inside attracts roaches and the pop bottle's shape keeps them from getting out, trapping them until they finally drown inside. Wine bottles, beer bottles and most long neck bottles will make good roach traps also. Always leave a little bit of the wine, beer, or soda in the bottle to help attract the roaches (or you can make your own Cockroach Brew, see this chapter).

You can add a little bit of soap to the left over soda, etc., in your soda bottle. Place the bottle in an area that they are seen, lean the bottle against something so that the roaches will have a way to climb in. Throw away the contents next day. Wide mouthed jars coated with petroleum jelly on the top 1 inch of the jar make excellent roach traps as well. Put Cockroach Brew into jar. Any thing like apple cider, wine, beer, sugar water, butter or dog food will attract them in and they cannot get out! Roach Bait stations are made with boric acid and a bait (attract silverfish too). A good roach trap is The Roach Eraser made by It Works Inc. .

Making Your Own Roach Sticky Trap: Place a piece of bread onto a board that has been spread with a layer of petroleum jelly, or Tangelfoot, or you can buy fly paper and use that. Rat trap sticky paper mats work well also. Another excellent roach trap is actually a flea trap (see flea chapter page). This flea trap also catches roaches on its sticky mat. I suggest that you add a small wad of peanut butter in the middle of the sticky mat. Replace sticky mat when roaches are stuck on it. Place in location where roaches are seen.

The Cockroach Inn: This unit is similar to the Ant Cafe or the Snail Inn. It is made from the same bird house (available at most pet stores) except it has a wider opening. Inside place a plastic cup. Screw down the lip. Inside the cup you can add the roach brew mixture. This unit keeps pets and kids out while allowing the cockroaches in.

STEP 4: WHAT DO YOU FEED THEM?

The Cockroach Brew

½ cup apple cider

½ cup cheap wine or beer or vinegar, apple cider

1 tablespoon of Brewers Yeast

1 tablespoonful of Butter

1 tablespoonful Dr. Bronners Peppermint soap

¼ tablespoonful of Boric acid * optional if children, animals around.

Mix every thing together into 8 oz. cup. Place into the Roach Inn. Screw down the lid. Keep children and pets out! Cockroaches will enter Roach Inn and go into the cup where the mixture is. Most die inside, others will die later. Change mixture and clean cup regularly. Keep count. Beer or wine, vinegar in a bowl, when used alone with brewer's yeast really works well in attracting and trapping them. Try soda also. Keeping count will tell you how well the trap is working and how much longer to continue using. This trap will attract the young roaches also.

Epsom Salts (Magnesium Sulfate) anyone?: Fill bowl one-half full with Epsom Salts. Place bowl in Roach Inns and place in area cockroaches are seen. Allow them access to the Epsom Salts by placing strips of wood against the bowl to allow the cockroaches to climb up. They will eat it. This will kill them within weeks. The magnesium will upset biologic systems in cockroaches and prevent them from feeding and they die.

2See also Hot Barriers page

3Remember use only natural soaps. See soap page

The Invisible Gardener Says: "Remember, Give yourself time to control roaches"

Homemade Containers: You can use wide mouth bottles, bowls, jars, whatever is available. I do not recommend this if you have children. Make sure it's child proof.

STEP 5: VACUUMING

Regular vacuuming of rugs, under furniture etc. will pick up egg cases. Dust with pyrethrum / DE mixture[4] ½ hr. before. Dispose of vacuum bags by placing inside plastic bag and tying shut.

STEP 6: GETTING A TUNE UP

Our Daily habits must be changed to create less favorable conditions for cockroaches to survive in. Think about what you are doing and how it is promoting cockroach life-styles. Controlling inside and outside water sources and food sources are two ways to insure that the roaches will not make your home their home too.

STEP 7: STORING FOOD

It is important not to provide food sources for them. There are many ways to store food. Using glass jars, canned goods. A seal a meal is a handy tool. Check your food containers. Throw away any food stuff with cockroach cases. Throw away into sealed plastic baggies. Roaches can transmit salmonella, boils, typhus, dysentery to name a few. Never leave food open over night. Remove stacks of newspaper, magazines, old books and grocery bags. Look for good hiding places and remove them.

STEP 8: WASTE MANAGEMENT

Throw kitchen wastes into composter. Use large sealable trash cans. Spray Dr. Bronners Peppermint soap around trash cans and on any roaches you see.

STEP 9: A LITTLE HANDY WORK

Keeping cockroaches out can be as simple as using rubber caulking around any entrances. Check around windows, doors, crevices, cracks, air vents, etc.. Check under house for possible entrances into house. Follow pipes into house. Check around connections, etc. Look for possible water sources. Repairs to the house are an important part of roach control.

STEP 10: MODIFYING THE ENVIRONMENT

The type of house we live in determines conditions that can lead to cockroach infestations. If you happen to live in a housing situation that promotes roach infestations, the first choice is to move. If this is not available then attempt to keep your immediate environment as clean and as healthy as possible to live in. Keep your home clean and let lots of light in. Another way to modify the environment is to use incense. Try varieties that will keep them away.

STEP 11: CLEANING UP

This is very important. If your environment feels good to you and encourages you to do better, then you will do better, and you will be able to control the cockroaches. Cockroaches have a well developed sense of smell. They will find any food that you leave for them that is out of reach of the ants.

---

4 Unlike dusting for fleas, you should dust only in areas where the roaches are seen. Avoid dusting rugs, etc. as this mixture might cause damage to the rug. Test first.

9

Natural Fly Control

Step 1: Control the Trash

Control fly population through proper disposal of kitchen and other wastes, including recyclable wastes. You can also control them through pickup & disposal of your pet's feces. Keep flies out of the house by using screens, bead curtains. Keep windows and screens in order. Use fly swatters. Use fly paper. There are many different types of stick-em's that you can hang up that will trap them. Reduce the sources of infestation and you will have a good hold on the fly population.

Step 2: Using Soap to Control Flies

Use soap to spray directly on flies and to spray on manure. Use either Dr. Bronners Peppermint soap or Jungle Rain. Use 1 cup per gallon water. Mist area where flies are seen. Great for using inside the house. Just mist the area and allow the flies to fly through the mist. This also wets their wings and allows you to swat them.

USING SOAP AND TABASCO SAUCE:

Use 1 tablespoon natural soap and 1 tablespoon Tabasco Sauce per gallon water. Spray directly on places where flies can lay their eggs. Not for use inside as the Tabasco will stain.

Step 3: Using a Natural Dust to Control Flies

USING DE

DE (garden grade...2) can be dusted on the manure to control flies. Also DE can be fed to your animals, horses. How to use DE is explained in other parts of this book. For cats feed 1/2 teaspoon per week. For dogs feed 1/2 tablespoon per week. For horses feed 1/4 cup per week added to their meals. The DE when eaten....3, will prevent flies from laying eggs in their manure. The DE will also help to de-worm as well as provide many trace minerals. See index for more info on using DE.

PYRETHRUM DUST

Dust around areas where flies are hanging out. Can be made into a spray by first making into a paste, slowly adding water to the pyrethrum, then adding that to water (see appendix for more info). There are many liquid pyrethrum products on the market, use only liquid pyrethrum that does not contain any additives. An excellent mixture is DE and Pyrethrum. Mix 50% See index for more info.

Step 4: Biological Control

Use parasitoids of flies such as Spalangia endius.

muscidiflurax zoraptor, Pachycrepoideus vindemiae, Tachinaepheus zaelandicus. Use beetles, mites or soldier bugs. When using parasitoids, do not use any sprays as it would hurt them also. Do not use DE either. Best to leave them alone to do their job. These good guys are available from Arbico, Peaceful Valley, as well as from most mail order catalogs. Fly parasites are small nocturnal burrowing insects. They do not harm mammals. They reproduce in three weeks depending on climate. They are shipped as parasitized pupae in a sawdust medium. Recommended is to release a small amount in places where flies are bad say around horse stables, manure dumps etc. They should be buried into the ground a few inches deep, and should be mulched over as protection. Should be done every year.

Use beetles, mites or soldier bugs. Fly parasites are small nocturnal burrowing insects. They do not harm mammals. They reproduce in three weeks depending on climate.

Step 5 Attractants and Traps

The Fly Terminator

This product has been around for a long time. Effectively attract and kills flies without poison scatter baits or electrocution. Uses attractants. This is a large bottle-like product that holds water. An attractant such as a piece of fish or fish emulsion can be used instead. The instructions say that no meat or fish is necessary because of the attractant they sell with it, however I found it preferable to use fish heads, etc., and/or to add a small amount of fish emulsion to the water and to change it every month or so as needed. The unit is placed or hung near where there is a fly problem. I have found this system to really work in attracting, catching, and killing flies. Available from most garden centers or through mail firms like Arbico, Gardeners Supply.

Fly Stick ems

There are many fly sticky tapes available on the market. Just hang em up! Available from Arbico, Peaceful Valley.

Peaceful Valley Fly Trap

A trap for catching large numbers of manure breeding flies. Trap will hold up to 25,000 flies. Non toxic yeast and ammonium carbonate bait attract the flies. Called Peaceful Valley Fly Trap because it's sold through them (Peaceful Valley Farm Supplies). Available only from (yes you guessed it) Peaceful Valley Farm Supplies!

PEACEFUL VALLEY FLY TRAP

Rescue Fly Trap

This is a handy disposable model. Just add water and hang. Throw away when full. Easy to use and easy to get. Available from most garden centers and through the mail.

Solar Fly Trap

An environmental safe aluminum trap that sits on the ground where flies dwell the most. Traps should be placed where they concentrate most, but not in the way of foot traffic. Bait is placed at bottom in a bait bowl. Flies are attracted by the bait, by pass it and move up the cone where they are trapped in the chamber and die in the sun. Sold by Arbico.

Organic Medfly Control

Medfly eradication efforts are proving to be a great waste of energy and money. Concentrating on the effect instead of the cause is wasted effort. Medfly spraying of Malathion or any other chemical is wasted effort and will not affect the real cause of the Medfly spreading in southern Calif. Medfly infestations are spreading throughout the world! Even this is not the real cause of the Medfly problem. The real cause is chemicals and the imbalances they have wreaked upon our planets eco-system. To regain balance we must stop relying on chemicals to control our pests but instead rely more on maintaining a balanced ecosystem, diversity rich with bacteria and enzymes. We must rely on sustainable organic methods of food production. We cannot depend on chemical control and the loop it causes forever. This system will break down and cause chaos and disorder, Re-Education of our farmers, retraining of professionals, re-teaching the teachers, is necessary. Our children must be taught the best way to grow and enjoy nature. Organics and the art of living without chemicals must be taught to all of our children and to their children and made into a way of life.

This is all well and good but what about now, what do we do now, you say?

Stop spraying any kind of chemicals. Start holding workshops to reach people on the various methods of natural Medfly control. Start workshops for farmers who can be affected by the Medfly. Show them natural methods of Medfly control. Encourage organic pest control methods with professionals by allowing for a natural/organic license for those wishing to use only organic methods of pest control. Encourage more research on developing natural predators. The best defense is a good offense. Provide for your property plenty of good rich compost, use only natural fertilizers and depend on natural predators to keep things under control. There are many natural predators of the Medfly. Many insects attack the fly and/or its eggs. Birds, toads, and other predators can keep fly populations down. Reduced use of chemicals increases functional diversity of life. It is this diversity that will maintain the balance. DE can be used to control the Medfly. DE can be added to water and sprayed on the fruit(s) affected by the Medfly. DE can also be sprayed on the ground around the fruit trees affected. Should be done in the spring.

THE FOLLOWING FORMULA CAN BE USED TO SPRAY ON MOST FRUITS AND VEGETABLES TO CONTROL VARIOUS FLIES, ESPECIALLY THE MEDFLY:

1 cup DE

20 drops natural soap such as Dr. Bronners Peppermint Soap or Jungle Rain.

5 drops Garlic Oil

10 tablespoon Tabasco Sauce.

Mix well in gallon container. Add water to fill. Stir well. Allow to settle and strain into sprayer. Spray on fruit and allow to dry. Can be sprayed directly on Medflies and other flies. Will also control flies and other insects so be careful where you spray.

"There are many natural predators of the Medfly. Many insects attack the fly and/or its eggs. Birds, toads, and other predators can keep fly populations down."

"Bats love to eat flies and other insects. Frogs love flies too and so many other creatures from lizards to birds! Provide homes for them so they will stick around!"

ANOTHER FORMULA

1 lb. compost

1 cup DE

1 cup molasses

soap

Place compost and DE into panty hose and place that into a gallon of water in a glass container. Add 1 cup molasses and allow to sit for three days in the sun. Add 20 drops soap. Add to sprayer and spray fruits, and other areas.

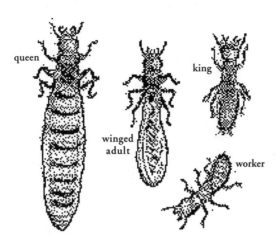

10

Natural Termite Control

While there are over several hundred varieties of termites in the world, only four live in the United States, subterranean, drywood, dampwood and powderpost. The subterranean being the one that causes 95% of the damage. Subterranean require contact with a moist soil. This limits their attack on wet and dry woods within reach of the soil. They make tunnel like tubes to connect the soil and wood. These tunnels are meant to protect them when they travel and are usually found free standing in the crawl space below your home. Drywood termites do not require the moisture that subterranean do. They can attack the structure far away from the soil. They create small galleries with small entrances plugged with partially chewed wood and a cement like secretion. Drywood termites leave piles of sawdust like pellets.

Termite Prevention

The conditions that invite wood damaging termites can be discovered and corrected long before the problem gets out of hand. Therefore it is important for the home owner to learn the proper measures that can be taken to prevent termites from entering your home.

Step 1   Inspection

POOR HOME MAINTENANCE

Repair loose or cracked siding or stucco, peeling paint and gaps around windows and doors that allow moisture into the wood that provides an attractive en-

vironment for them to live in. These exteriors should be properly repaired or replaced. Peeling paint should be removed and replaced. Adding boric acid to the paint will increase its effectiveness.

Low foundation walls that allow close to the earth to wood contact are the main areas to look out for as these are prime conditions for the termites to enter the house. The foundation should be raised or the earth lowered at least 6 inches preferably 12-18 inches. Repair all cracks in the foundation that provide access to the wood inside. Look out for the gaps at delaminating brick veneers which provide easy access to the wood below. Repair or replace. Must be water tight.

Poor ventilation in crawl spaces will encourage moisture and dampness will occur which is again great conditions for the termites to live in. Insure crawl space vents for proper ventilation of all areas. The vents can also be dusted to dust the area with. I will explain this later on dusting. Check for plant growth directly outside vents to insure that it's not inhibiting proper ventilation. Remove or trim any brush, etc., as needed. Always avoid plantings that are closer than 2 feet from the house. A good rule of thumb for crawl space ventilation is about two square feet of opening for every 25 lineal feet of wall.

FIREWOOD STORING

Avoid storing your fire wood or wood scraps next to the house as this could support termites and allow them entrance into the house. Also remove any old tree stumps which you may have around as this may become host for the termites.

STRUCTURE CONTROL

STRUCTURE CONTROL

Porches, decks, and other wooden structures should not be in direct contact with the soil. Concrete footings should be used. I suggest that you use builder's sand around the base of any wooden structures at least 2 feet out and two feet deep. Sub termites will not pass thru this barrier. I suggest that you mix DE with the builders sand as you apply it. See barrier's page.

PLANT CONTAINER CONTROL

Avoid plant boxes attached to the house unless it's off the ground. Control the watering to avoid constantly wet wood. Soil nematodes can be added to the planter boxes to discourage termites. See Using nematode's page.

WATER CONTROL

Leaking pipes or water faucets will keep the wood moist causing damage to the wood and providing access to the termites. Look for leaks. Check the water meter after turning off all water use. If still moving then you have a leak. Leak control is very important in controlling termites. Control your watering by way of drip or soaker systems. Over watered landscape is a major part in attracting termites. Provide proper compost and mulch to insure healthy balanced soil bacteria.

OTHER CONSIDERATIONS

Pay special attention to the bathroom, kitchen and laundry rooms as these are the places most of your water is used and a great number of problems arise. Check for leaks and repair or replace as needed.

Step 2: Diagnosis

Once you have found signs of wood damage and you have determined that it's not just old decaying wood then you have to decide if it's termites or carpenter ants that you have. There are also termite inspecting Dogs called TADD Dogs. If you have determined that its carpenter ants then go to the Dances with Ants chapter and treat like you would the ants, the only difference would be that you place the ant cafes inside the house, in the attic, crawl spaces, any place where they are seen, hidden from view.

Do it yourself: Aside from inspecting your home yourself for obvious signs of termite damage, there are many things you can do yourself:

Step 3: Methods of Control

DUSTING

Here are some natural dusts you can use around the house to control termites. Dust in places such as attic, crawl spaces, around outside house where wood meets the ground or other moisture.

DE: DE is one of the safest methods of providing termite control. See DE page in Appendix and in index for more info on DE. DE can also be mixed in with your builders sand for added protection. DE can also be

Diatomaceous Earth

"DE is one of the safest methods of providing termite control."

painted on the wood. I suggest 1 part DE to ¼ part Boric acid, added to water or to paint until a thick paint like texture, then painted on the wood.

Pyrethrum: Pyrethrum can be added to the DE and painted on for additional protection. Best used as a dust for immediate contact kill. Mix with equal parts DE for an excellent dust. See pyrethrum in index for more info.

Step 4: Spraying

Spraying of attics and other crawl spaces with a liquid DE/Boric acid mixture is highly recommended at least once per year in the early spring. Wear a face mask and avoid breathing. Use a Dustin Miser.

Step 5: Surgery

Removal of their colony and tubes is easily done upon detection. Remove infected wood and replace with newly treated wood (you treat yourself with the Boric acid/DE paint). Look for signs of their tubes and remove these. You can also dust the areas with DE.

Step 6: Boric Acid

Boric acid can be painted on to the wood as a form of protection against termites eating into it. You can also use 20 mule team borax to dust into their tunnels. Avoid breathing. See Index for more info on Boric acid.

Step 7: Barriers

Construction Sand (90%) and DE (10%) mixed together and laid down around the house 3 feet deep and 3 feet out from the house will prevent termites from moving into your house.

Step 8: Nematodes

Termask (Steinernema carpocapsae), are termite eating nematodes available thru Arbico and many mail order companies. They are easy to apply (using a garden hose and attachment unit) and are safe as well. Should be used on a yearly bases to insure effectiveness. Works only on subterranean termites not drywood.

Your Ant Friends

Termites and most ants are enemies and will fight each other. Carpenter ants are often mistaken for termites (see below) and must be avoided but most other ants can be allowed to live outside (provided they do not come in (see ants section for proper control). Which would you rather have?

Step 9: For Hire

If you have done the above and still have termites then I would suggest this section.

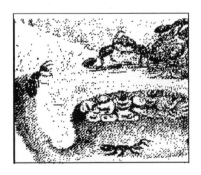

Termites and most ants are enemies and will fight each other.

CARPENTER ANT

"Carpenter ants and termites both behave the same way. That is to say they both live in wood, both have wings during mating seasons and they both leave signs(piles of wood dust, wings) which lets us know they are there. Can you tell the difference?"

TERMITE

COMMERCIAL METHODS AVAILABLE FOR TERMITE CONTROL

<u>Heat:</u> In which propane heat in a tarp wrapped house is used to raise the temperature of the wood to 120 degrees (room air to 187 degrees) This method is called Thermal Pest Eradication.

Effective rate: 95%

Limitations: for best results should be done every 5 years (more often in tropical climates).

Best combined with another type of treatment such as Liquid nitrogen for spot treatments.

<u>Liquid Nitrogen:</u> Liquid nitrogen, 292 degrees below zero, is sprayed into walls and pinholes.

Effective rate: 100%

Limitations: cannot reach all parts of the house, best combined with another type of treatment such as electricity. The Tallon company in California uses a system called The Blizzard.

<u>Electricity:</u> Using an Electrogun, electric shock (90 watts - 90,000 volts) is applied to the walls and other areas of the house thru tiny pinholes that are drilled and repaired after use. Penetrates up to 1 to 2 inches into the wood. Nails can be used to extend its reach.

Effective rate: 80%

Limitations: Cannot reach all parts of the house, best combined with either liquid nitrogen or heat or micro wave. Used as a spot treatment.

<u>Microwave:</u> 500 watts is applied to the house thru a funneled device the size of a toaster oven setup on a tripod close to the wall.

Effective rate: 90%

Limitations: Cannot reach all parts of the house, best combined with another system such as heat or electricity.

ABOUT CARPENTER ANTS

Carpenter ants and termites both behave the same way. That is to say they both live in wood, both have wings during mating seasons and they both leave signs (piles of wood dust, wings) which lets us know they are there. Can you tell the difference? Here are a few of their differences:

ANTS	TERMITES
Antennae elbowed	Antennae not elbowed
Two pairs of wings of unequal length	Equal length wings
Eyes visible	No eyes visible
Thin waist	Thick waist

11

Natural Lawn Care

The dangers to our health from the chemicals being used to keep our lawns green and weed free are increasingly becoming more apparent. Strangely named chemicals such as pendimethalin, benomyl and 2-methylcydohexy are widely used to combat weeds. There is an ever growing list of chemicals used to maintain a "healthy lawn" along with the use of chemical fertilizers, herbicides, pesticides and who knows what else, causing concern amongst the public. These chemicals not only have dire effects on the environment but on we 'humans' as well. We can no longer trust the judgment of the EPA.

"Being registered with the EPA by no means counts for a seal of approval or a seal of safety", says the National Coalition Against the Misuse of Pesticides.

Therefore, it is important to begin to take it upon ourselves to control the chemicals we use in our everyday lives, such as the chemicals we use on our lawns. This chapter is about how to grow a deep green healthy lawn without the use of man made fertilizers and pesticides. We will discuss Preventive Cultural Practices that is the foundation a healthy lush green lawn is based upon. Understanding the components of the lawn's ecosystem and providing a more suitable environment for the types of grasses and microorganisms that grow best there while decreasing the conditions for pests and harmful bacteria's, is the "ideal" and more natural, less toxic method of providing yourself with a beautiful lawn.

Preventative Cultural Practices

PCP includes the proper selection of the types of grasses for your area and the lawns use (are you going to play on it, walk on it a lot, or just enjoy its beauty)? Having a healthy lawn without harsh chemicals is really very easy to do. Lets go over the basic steps together.

Step 1 Understanding The Soil

When you understand the type of soil in which your lawn is to grow or is already growing, then you can begin to create a soil structure that is better suited for proper plant growth. Soil supplies air, water, nutrients, and physical support to plants. Soil also needs organic matter to keep it alive. Each type of soil needs a different amendment. Sand, clay, and silt, in varying amounts, determine the texture of the soil. Compost will help to bind soil particles and hold water to the soil. Water is also important for productivity. Too much or too little water hinders proper plant growth. In addition, the presence of earthworms aids in maintaining good soil structure. Use only organic materials. Increase organic matter. Use manure as a top dressing. Use organic fertilizers. Lawns love compost[1]. Lawns love rock dust. Blend the two for best results.

Chemical fertilizers kill off the beneficial soil bacteria, as well as killing off earthworms and therefore should not be used, should be avoided at all cost! In this environment, the lawn and its bio-system will be

---

1 See rock dust chapter for more info.

operating under stress. Stress is the most important factor in pest control. Whenever the lawn is under stress, several things have occurred: **1:** Proper Food is not being made available to the lawn. **2:** The Humus levels have dropped below optimum. **3:** Water effects stress. Too much or too little water will cause stress in the lawn (chlorinated water is bad for the soil). 4: High nitrogen fertilizers kill the soil and cause a great deal of stress. Avoid using Urea at all costs.

UNDERSTANDING THE ENVIRONMENT

More than any other element of climate, temperature will determine which type of lawn will grow where. Available sunshine and water (either as rain or humidity) round things out. An example of this is: if irrigation is unavailable, only certain desert species of lawns can be grown. It is a good idea to consult a gardening book that has a map broken into climactic zones. This will help to simplify the process of deciding which grasses you should consider.

Step 2   Soil Amendments

There are many forms of organic matter. Untreated sawdust, aged horse manure, and aged wood are common amendments and can be found anywhere. Aged wood soil conditioner, cocoa bean, mushroom compost, rice hulls or apple pomace are usually available regionally. Organic waste from your kitchen can be used to compost your soil as well as additions of rock dust. Other amendments are: leaf mold, pine bark and bark chips, straw, sand, coffee grounds, grass clippings, shredded cardboard, bat guano, eggshells, grapefruit skins, potato skins, and wood chips and ashes. Organic matter should be added to your soil before you start a new lawn. Should be done when the weather is fairly settled, so the ground will be prime for planting and the new grass will have only mild competition from weeds. The Best Organic matter to add is as Compost. See chapter on compost.

ORGANIC FERTILIZERS

Soil, in order to be alive, must have high organic matter, drainage and good structure. Natural fertilizers add to the soil, improving its fertility, maintaining and contributing to the improvement of these necessary elements. Organic matter and rock powders form the basis of organic fertilizers, and benefit the soil as well as the plant. As a rule, chemical fertilizers are not a complete plant food. Organic material contains nutrients that provide the microorganisms in the soil with the materials they need to be active.

Organic Fertilizer mixes usually contain composted animal manure, plant residues, seaweed and fish products, and minerals (bone and blood meal, cottonseed meal, granite dust, phosphate rock and greensand). Organic Fertilizers should be used in combination with compost in order to develop rich, humusy soil. Compost: Composting is practiced today just as it was hundreds or thousands of years ago. The recirculation of dead matter into life is a part of nature's program of soil rejuvenation. A gardener's compost heap is a process that is going on eternally in nature. When we cut the grass and remove it we are cutting off the cycle. Therefore compost, when added back to the lawn, reestablishes the cycle and returns nutrients

"There are two ways to sterilize the soil; by Heat and Chemically"

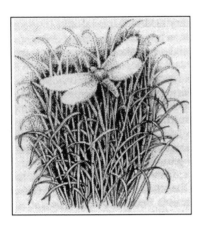

"Compost is the Finest Natural Fertilizer there is"

and bacteria to the soil. The gardener takes a tip from nature and uses this simple method to build the soil's fertility.

Horse Manure

Among the manure of other farm animals, horse manure is one of the most valuable. It is richer in nitrogen than either cow or hog manure, and ferments much more quickly, therefore being referred to as "hot manure". Manure provides organic matter and trace minerals to the soil. Your lawn will benefit by having a thin layer of aged (at least 6 months old) horse manure spread over it (using a manure spreader), and then watered well. This should be done twice per year or even seasonally if you can do it.

This practice will also benefit the microbes as well as the beneficial insects. But the best top dressings are made from well made compost because it provides a more complete and varied food source as well as varied bacterial sources. Run through shredder for best results.

Step 3   Watering Procedures

Deep, regular waterings are essential to develop a lush, verdant lawn. Lawn sprinklers are a definite improvement over hand watering. They can be set in place to water as long as necessary, and are especially effective if placed on a timer. The amount of water used depends on how deep the roots are of the grass you are growing. If there is a drought, it is better, rather than sprinkling lightly, to water twice a week, soaking the soil to a minimum of 4 inches, or not at all, for shallow watering causes roots to spread out near the top where they will be baked by the sun, becoming unable to withstand drought. A lawn that is watered thoroughly at regular intervals, and whose soil has plenty of organic humus (organic matter converts into humus that makes the nutrient elements in the soil available to the grass), will withstand drought, and remain sound throughout the hot summer months. Different types of soil need different amounts of water. Light soils need more water because they drain so rapidly. Clay soils retain water, so they need less.

Step 4   Pest Identification and Tolerance levels:

Not all pests are bad for the lawn in low levels for they perform a desired task in maintaining a balanced ecosystem. This applies to weeds as well as insects. 5 to 10% weed growth is allowable and often not noticed. This also depends on whether you are offended by a certain type of weed and what weeds do not bother you. This will make a difference to you as to the over all appearance of the lawn. Decide what levels you will tolerate and operate within them. As for pests, develop a monitoring system that will allow you to determine what steps are needed to keep these pests below your tolerance level. Learn to identify the good guys from the bad guys. This is important! There are much good bug ID books out there.

Step 5   Monitoring Systems

Monitoring is when you regularly inspect the lawn for signs of harmful and beneficial lawn pests. Before you cut the lawn, you should inspect it for signs of stress.   Here a garden notebook becomes very handy. Write the date and location and what signs you have observed and write down what you did to correct the problem.

A lawn that is watered thoroughly at regular intervals, and whose soil has plenty of organic humus (organic matter converts into humus that makes the nutrient elements in the soil available to the grass), will withstand drought, and remain sound throughout the hot summer months.

"Your lawn will benefit by having a thin layer of aged horse manure spread over it."

Step 6 Developing a regular maintenance program

It is important that you decide what work must be done on what basis, and do it. Watering, fertilization, pest control, and maintenance should all be done on a regular basis, and notes should be kept on the whole process.

Here are some maintenance steps that you should follow:

☑ Soil Test... once per year. See Green Gem Appendix 9 page 166

☑ Organic Fertilization... twice per year.

☑ Top Dressing of lawn with compost... every season.

☑ Irrigation... establish a regular program of deep waterings on the same days at least two or three times per week. Watering early in the morning is usually best.

☑ Aeration depends on soil structure and should be done once per year.

☑ Maintenance... Cutting the grass is a very important step. Cutting too short or too long depends upon the type of lawn and the type of lawn mower.

☑ The Mulching Mower

☑ Using a mulching mower is one of the best ways to recycle your grass clippings and feed the lawn at the same time! Try the newer solar mulching mowers!

Step 7 Identifying Lawn Problems

PESTS OF THE LAWN

Caterpillars/Grubs: Caterpillars/Grubs are found in the soil. Grubs cause damage to the lawn by leaving yellow or brown patches (roots of lawns are eaten). More pests are listed on website.

Soap Drenching: You can check for grubs by placing a can that has be cut at both ends, place one end into the lawn then pour soapy water into it. In about 10-15 min., see if any grubs have come to the surface. These grubs can be dealt with, by using beneficial nematodes. Just follow instructions for its application. In most cases it's a simple matter of dissolving in water then spraying on areas. Compost applications and avoidance of high nitrogen fertilizers will help to control in future. Beneficial nematodes are highly recommended for this control. Try Arbico. Japanese Beetles love poor dead soil since their natural enemies are not present. The Grubs of these are a very serious threat to most lawns. A natural lawn will not be attacked by the beetles as an over chemicalized lawn would be. Beneficial nematodes provide excellent control. Milky spore disease is highly effective against this and most beetles.

Chinch bugs are also serious pests of over chemicalized lawns. Use the soapy soil drench method as described above to see if any chinch bugs come to the surface. If you think your lawn has chinch annoys then you must first pay attention to proper caring for your lawn. Follow the basic steps mentioned in this chapter for organic lawn care. Top dressing of compost is a necessity.

A Natural Chinch Bug Spray

Mix together Jungle Rain, and Dr. Bronners Peppermint Soap. Mix equal amounts of each. To a gallon sprayer add ¼ cup and spray on the lawn. Best to spray in late afternoon to avoid any damage. Do not spray during a hot day. Water in well afterwards.

Chinch Bug

Moles in the lawn: See the chapter on gopher control and follow same procedure.

Gophers in the lawn: See chapter on gopher control.

Ants in the lawn: see chapter on Ant control. Follow the procedures described in the ant chapter but also use the following on your lawn.

Steps to remove ant hills from your lawn:

1. To one gallon water add 1 cup soap and pour into ant hills.

2. Mix equal amounts of DE and Pyrethrum dust. Dust hills, turn over with shovel to uncover ant colonies.

3. As last resort pour boiling water into ant hills. See ant chapter for more info on ants.

Mining Bees in the lawn: Their mounds can be removed with a rake. They do very little damage to the lawn.

Dogs peeing on the lawn: Hose with water as soon as possible. Top dress with compost.

Step 8 Lawn Damage

Drought Damage is caused due to improper water and composting of your lawn. Watering deeply less often is best. Aerate the soil to allow water to enter. Chemical spills can be cleaned up with a little citrus soap and water. Hose down well after wards. Compaction is due to improper aeration and watering. Top dress with lots of compost or old horse manure once per year. Fertilizer burns can be helped with lots of watering to flush away chemicals. Top dress with compost to allow lawn to come back. Pesticide burns should be treated the same as above.

Step 9 Diseases of the Lawn

Alga's growth generally does very little damage to lawns. Main cause is over-watering and high nitrogen fertilizers. Top dressing with compost will help conditions.

Fungi Diseases: Fungal Diseases are caused by dead soil and using high nitrogen fertilizers. Over chemical use on lawns eventually will kill off all the beneficial bacteria in the soil, leaving proper conditions for the bad guys to move in. Best type of fertilizer for the lawn is properly made compost, or other natural fertilizers such as Rock dust, Nature meal (Nitron), and many other natural lawn fertilizers available on the market today. Use Nitron A-35 as a bacterial activator (also try Agri-Gro, SuperSeaweed, Acadie, etc.). Wettable sulfur makes a good immediate fungal control.

Lichen growth can be removed by hand and raked clean. Top dress lawn with good rich compost and spray nutrition such as Superseaweed, Agri-Gro, Nitron A-35, Fish emulsion, etc. Toadstools or mushrooms are not necessarily to be viewed as bad for the lawn but instead it should be looked upon as an indication that the conditions of the lawn should be looked at more closely. If your lawn has toadstools then pick them and dispose of. Top dress the lawn with good rich compost. Keep an eye on your watering. Water in early AM is best.

"Disease can be avoided by proper fertilization and proper watering techniques"

Making your own Organic Lawn Fertilizer

5 lbs alfalfa meal

5 lbs rock dust or gypsum

10 lbs compost (finely grounded)

10 lbs aged chicken, llama or rabbit manure (finely ground)

5 lbs coffee grounds

5 lbs white sugar

Mix well together. All items should be run thru a fine screen. Apply as needed.

Making your own liquid lawn food for spraying

1 cup Nitron A-35

1 capful Superseaweed

1 cup Agri-Gro

¼ lb. rock dust

¼ lb. afalfa meal

1 cup liquid seaweed like Acadie

1 cup molasses or sugar

1 can of Beer²

Mix together and place into a panty hose tied into a ball. Allow to sit in filtered water or solarized water for a day (add liquids before), strain and spray onto lawns. Best time is early am. See making your own foliar spray chapter for more.

Step 10   Natural Weed Control

Weeds will only grow in soil that is not rich in minerals. Weeds need minerally deficient soil to grow in, so keep your lawn well fed with minerals and natural bacteria. You will have very few weeds establishing themselves in your lawn.

Proper weed control is established thru:

PROPER MOWING OF LAWNS

Always pull up weeds before you mow to avoid spreading the seeds. Keep a sharp blade.

CORRECT WATERING

Regular deep watering is better then often shallow watering. Watering in AM is better then watering in PM.

CORRECT FERTILIZATION

Stop using chemical fertilizers. Stop the high nitrogen cycle. Organics will provide plenty of nitrogen etc. as needed by the lawn for healthy weed free growth. Give lawns plenty of rock dust that provides minerals needed by healthy lawns. Provide your lawns with a top dressing of old horse manure or aged compost at least yearly. Chicken manure top dressed on your lawn 4 times per year will keep your lawn green and healthy. There are many natural fertilizers available on the market today.

Some Places Organic Lawn Fertilizers are available: Nitron Industries, Arbico, Gardeners Supply, Albright Seed company, C.P. Organics. See also local nurseries, see also mail order catalogs.

HAND WEEDING

Hand pulling of the weeds is an easy and natural way to control weeds. I suggest that you first spray the weeds with a mixture of natural soap and Safer Insecticidal soap (50/50) the day before (see below). This makes pulling the weeds up easier. Weeds can be controlled by using natural soaps. Soap will burn them and they will die without having to remove them from the soil. They will decompose and provide food for the soil. Test the strength of the soap mixture. Make to strength as needed.

Use any natural soap like Dr. Bronners peppermint soap and Jungle Rain. Mix these two soaps together at 50/50 ratio. Add 1 cup of this mixture into a gallon sprayer. Should work on most weeds but will not damage the lawn. Do not water for 24 hr. and do not spray in midday or on a hot day.

---

2 Try to make your own lawn beer. Or buy the type of beer with the lest additives and preservatives. etc.

12

Organic Rose Care

Step 1 Getting Started

The first thing you have to do is address the nutritional needs of your property as well as that of the plants. Proper nutrition is the KEY to natural pest control. Anytime you treat the cause, you will be controlling its effects. Ask yourself "what am I doing to the soil"? Stop using chemical fertilizers. Stop using pesticides that are not made naturally. Avoid high nitrogen at all costs.

Yes, you will have to switch to organics if you want to regularly control thrips, aphids, and other pests naturally. Organics and chemicals do not mix well. You may get away with good results for a few years longer then if you were completely chemical but sooner or later you will pay for it, in the form of one pest or another, or one disease or another, not just the plants and soil suffer; you will also suffer from exposure to the chemicals in the fertilizer or pesticide, etc.

Fertilizer exposure can be as dangerous as pesticide exposure. Remember, you cannot be organic only when it suits you to do so. Organic demands it to be your way of life.

Take Good Care of the Soil and the Soil will take good care of your Roses!

If you are a first time rose grower there are many things we must agree upon before starting. How much time do you plan to spend on your roses? Are you a part time rose lover or are you going to really go at it? There

are many good books available on the subject of growing roses. Take an organic rose growing class. Growing roses for show is a great deal of fun and lots of work. You will also have to decide if you are growing organically or if you will be using chemicals.

ORGANIC OR CHEMICAL ?

Our health is the number one reason for growing Roses organically. Consider the many chemicals on the market today and the many health problems that arise from their continual use. Remember that children and older adults are at greater risk. Furthermore, consider the environmental pollution associated with their production, transport and consumer use. You can have that perfect show rose grown organically and stay healthy to enjoy it. An added benefit is that roses organically grown are more disease resistant, & they last longer!

THE BIG SWITCH

Going from Chemical to Organic can be a traumatic experience to both the rose and the grower. Not only Plants but people go through chemical withdrawals. You must allow at least one year to complete the switch (see converting to Organics). This allows both you and the plant time to adjust to the new organic regime. While it is best to start off organically, you may not be able to. Commercial rose growers are not organic and therefore when you first buy your roses they have been grown chemically! Don't Panic! Given time your roses will love being grown organically.

WHICH VARIETY?

The varieties you choose are important. Environmental conditions determine what varieties will do well and what will not. Don't just plant any rose or you will be sorry! Consider the following: Is it hot and dry? Is it very wet? Fog? Winds? Where are you located? Your state will have its own special environmental issues that you will have to deal with. What varieties are being grown in your location?

LOCATION

Choose an airy location to allow for air circulation to reduce fungus development. A sunny location is necessary if you are going to have stress free roses. Early morning sun is best for roses. They like to look at sunrises! Give them as much sunlight as possible.

SOIL CONDITION

The soil's condition determines everything. If you have all the other conditions correct and not the soil, you will not be able to grow them organically and chemicals will be needed. The key to growing anything organically is healthy soil. Living soil allows for a functional diversity of organisms. Functional diversity is how organisms cooperate with each other. This sharing of resources provides for an ecologically balanced system. The greater the balance the higher the energy level, the greater the nutritional levels, the less the stress and the less the pest.

KNOW YOUR SOIL PH

Different plants require a different PH for maximum benefit. Know your plants and their PH requirements. In soil where the PH is too high it will not be able to assimilate certain minerals resulting in a trace mineral deficiency that invites a pest or disease attack. Most animal manure's have a high PH and you must use peat moss, forest shavings, etc., to counter balance this. Rock dust also has a high PH (8-9 PH) and therefore must be balanced with a slightly acid mulch/soil, compost. Roses love PH of 6.5

Step 2 Watering
SLOW AND DEEP

If you have frogs hopping around your roses you probably are watering too much! Roses do not like overhead watering. Always provide either a soaker hose or a drip system for them. Regular watering is important. Avoid over watering. Slow deep watering is the best. Use a 2 gallon per hour drip head, one on either side at approx. 6 inches out. A soaker hose will work fine for this purpose as long as it does not spray water onto the leaves. A soaker hose can be buried or mulched over to prevent this. Dig a well around the rose to hold in the water and the compost and mulch. Mulch should be 2-3 inches deep.

Step 3 Nutrition
COMPOST AND ROSES

Get good at making compost. Experience is the best teacher here. The better your compost, the better it will be for your plants and the more effective the organic system will be. Remember, proper nutrition is the

Take Good Care of the Soil and the Soil will take good care of your Roses!

If you have frogs hopping around your roses you probably are watering too much! Roses do not like overhead watering. Always provide either a soaker hose or a drip system for them.

corner stone of the organic system and compost is its main building block. Roses love compost so feed it a cup of good rich compost once per month.

FERTILIZING YOUR ROSES

Avoid using high nitrogen fertilizers: Nitrogen is naturally provided in the organic system and is never lacking. In the organic system, nitrogen is easily available and only used when needed and in small amounts for longer lasting results. Some natural sources of nitrogen are animal manure's such as horse, cattle, llama, rabbit, earthworm casting, chicken etc.; they also provide natural bacteria, enzymes, and trace minerals.

Organic Fertilizers: Chemical fertilizers lack bacteria and enzymes essential for soil life and for nutritional exchanges necessary between plant and soil. Compost is the finest organic fertilizer you can use on your roses. However, you will need to add a good organic fertilizer, approx. 1 cup per month per plant. The Organic fertilizer should be a 5-5-5 with Nitrogen no higher than 7. There are many good organic fertilizers on the market today. Read the ingredients. Avoid urea base fertilizers. Find two or more different organic fertilizers and switch between them. Or you can make your fertilizer.

Minerals: Minerals in a fine form such as rock dust, granite dust, decomposed gravel, greensand, soft rock phosphate. Minerals from the ocean such as kelp meal, fish meal, seaweed powder, crushed oyster shells. Minerals from the animal kingdom such as bone meal, feather meal, minerals from animal manure.

Using Rock Dust: Rock dust will help to increase the energy level of the soil and in turn will quickly raise the roses and the soils energy levels. This is primarily due to its high calcium levels as well as high iron, and its large selection of trace minerals, which are made immediately available to the soil and plants. To avoid the dust, rock dust can be made into a milk like liquid and sprayed on the leaves. Use only 1 tablespoonful of rock dust and 1 tablespoon of Diatomaceous Earth (garden grade only) per gallon of distilled or solarized water (Stir in a clockwise direction for 1 min. then quickly stirring in the opposite direction for an additional min.). Allow to settle for 5 min. Add 10 drops Superseaweed™ or any concentrated liquid seaweed or natural fish emulsion. Strain into sprayer. Spray once per month, or daily, as long as pest infestation occurs.

Step 4; It's Alive!

BACTERIAL SPRAYS

Bacteria are nature's cooks. They take raw materials (minerals) and eat them, converting the minerals into compounds that are easily assimilated by the plants. The bacteria eat first! We are in essence feeding the soil! Try to use a garden transformer to remove the chlorine in the city water. Chlorine kills bacteria. YBM makes the only water transformer I have found on the market for this purpose. See Index and appendix.

For Bacterial Sprays use any of the following:

Nitron A-35 tm: A Bacteriological activator. Provides enzymes. Helps bring the soil back to life. Use at 1 cup per plant per month. OK to use along with organic fertilizer but will need to use ¼ cup fertilizer less, as the

Making your Own Organic Rose Fertilizer

1 lb. New Jersey Greensand

2 lbs Rock Dust

2 lbs Alfalfa meal (organic)

2 lbs Fish meal

2 lbs Seaweed powder like Acadie, etc.

2 lbs Earthworm Castings

½ lb. Epsom Salt

1 lb. Chelated Iron

Mix together, use at 1 cup per plant per month. Water well. See also organic fertilizer chapter for another version of rose fertilizer.

Some More Bacterial Sources: **Animal Manures** provide many different types of bacteria needed by the soil. Choose from different types of animal manures. Horse, cattle, sheep, llama, rabbit, are some to name a few. Do not use your dog or cat manure. Do not use any carnivorous animals as their manure could contain harmful parasites. Take a drive out into the country and discover for yourself what resources are available for you in that area. Chicken farms, (egg farms), rabbit farms, dairies, horse farms, etc. all make good sources of these materials. Talk with them and see what they want for it. Sometimes they will give it to you free! Ask them if they spray any chemicals on the manure. Avoid using any manure that has been recently sprayed. **Allow at least 6 months before using.** Composting will remove most chemicals.

Nitron will make more of the fertilizer available. See index.

Agri-Grotm: Provides many natural bacteria needed by the soil. See foliar spray chapter for more info and index.

Using a Liquid Seaweed: Seaweed is full of trace mineral and bacteria. A good seaweed is Acadie. See index.

SuperSeaweedtm: Provides trace minerals, bacteria. A natural Biodynamic spray. Use 10 drops per gallon. For best results add 10 drops per gallon of filtered water one day before use and allow to sit. Superseaweed can be added to Nitron for even better results. For more information see chapter on Making your Superseaweed. Available at IG of A. See index.

Compost Tea: One of the best bacterial sprays is made from compost tea. Place 1 cup of compost into a panty hose and tie into a ball. Suspend in a glass gallon container filled with filtered water and add 1 cup molasses. Allow to sit in sun for 24 hours. Spray in evening. A good program is to add 1 cup Nitron, 1 cup Agrigro, 10 drops Superseaweed, 1 cup Acadie, into gallon of compost tea. Allow to sit in sun for 24 hours, then spray on roses. See Compost Tea in index for more info.

Milk: An excellent bacterial spray, milk is also a good source of calcium. Milk provides natural bacteria that prevents and fights various diseases. Milk will kill off fungal diseases while allowing beneficial bacteria to grow. Fill a gallon sprayer to within 3 inches of the top with filtered water or solarized water. Solarized water can be made by filling a gallon glass container. (A good container is the water companies 5 gallon bottle.)[1]. Allow to sit in sun. Shake regularly to remove any chlorine, etc. One or two days will be long enough. Spray on leaves. You can add a liquid seaweed concentrate such as Superseaweed™ or Acadie™ to increase effectiveness.

Fish Emulsion: Fish emulsion makes a great bacterial spray. Try deodorized fish emulsion for less smell. Avoid fish emulsion that has urea in it. Nitron makes the only fish emulsion without Urea (that I know of) so get that. See index.

Organic Alfalfa Tea: The biodynamic farmers knew the benefits of spraying with Alfalfa tea in early spring. Alfalfa is very high in nitrogen (10%), very high in many trace minerals, high in iron and especially high in natural bacteria. To make Alfalfa tea add 5 cups Alfalfa meal into a 5 gallon container of filtered water. Allow to sit for one day. Then add ½ cup fish emulsion (without UREA), 1 cup Nitron A-35, 1 cup Agri-Gro, 1 cup Acadie, 20 drops Superseaweed and 1 cup rock dust. Stir well and allow to sit in sun for another day. Should start to smell just right. Add 1 cup of this liquid to ½ gallon of water. Can be poured around base of roses or sprayed on leaves. If spraying, filter before pouring into sprayer. You can get Alfalfa meal from Nitron, Arbico, Peaceful Valley.

Step 5 Mulch Well

Mulching is very important. Use acid mulch if soil is too alkaline, use alkaline mulch if soil is too acid. Mulching helps retain moisture, keeps nutrition levels high and reduces stress. Proper mulching increases biodiversity. The more bio-diversity, the greater benefits between living organisms. This reduces any infestation of any one species (like snail control).

Diseases and Pests of Roses

Once you have begun to provide for a proper nutritional environment, you must provide protection until balance has been regained. You can protect your roses from thrips, aphids and other pest attacks thru several different methods: Planting garlic at base of rose will deter pest attacks. Society garlic is very effective. You can also make a garlic oil spray. Buy garlic oil from your local grocery store. If they do not have garlic oil buy garlic butter. Add 1 tablespoon to a quart water sprayer. Make sure to strain before pouring into sprayer to prevent clogging.

Another method you can use to control pests on your roses is to bury tobacco around the base of your roses. For best results take 1 cup dried tobacco (use only organically grown tobacco if possible), ½ cup garlic powder and mix into this 1 cup compost. Bury at base of rose by turning over into the soil. A better method and one that I prefer, is to get a small clay drain pipe (opened at both ends, about 3 inches wide, 12 inches long (smaller for roses, larger for trees) and bury that at the base of the rose bush or plant, level with the ground. Into this place the mixture TGC+™ (equal amounts of organic tobacco, garlic powder, compost and the + stands for trace minerals such as rock dust). Use ½ lb. per unit. On top of this place either a pretty rock or a nice clay pot to cover it, or simply mulch over, remembering where it is. This mixture should be replaced once per year or as often as needed.

You can also use pebbles as a mulch to cover it, just remember where it is. For this rose vent to be effective, there should be a drip head (2 gallons per hour) to allow water to pass thru it. The TGC+ will be absorbed by the rose, Ficus etc., and it will kill the thrips, aphids, etc., as they attack the rose leaves (anything that eats these leaves will die!). Tobacco and garlic are both absorbed into the plant, anything that attacks the plant will also get this mixture. The tobacco will kill any insect that attacks it. The Tobacco is very volatile and will biodegrade within 24 hr. if in liquid form; will last longer if it is in a dust or in its natural leaf state. Do not use tobacco on fruit trees. The garlic will change the taste of the plant confusing the bug.

Controlling Aphids and Ants on Roses

The other day I had a caller on my radio show. She had heard my previous remarks to a caller concerning ants and aphids and their relationship. This caller was certain that only some aphids are herded by ants and that most fly from place to place totally indifferent to what the ants have to say about it. It is just the opposite, that most aphids are under the control of their ant masters. I also have found it to be true that by controlling the ants, you obtain a greater control of the aphids that are attacking your roses, etc. There is a definite relationship between various insects, in particular aphids, which are found on plants, and ants. The ants control most if not all insect activity on plants that they have 'adopted'.

---

1 Please note: There are many different products on the markets that can be used in place of the above. We however can only recommend those products that we have used and tested in our organic testing gardens. If you have a product(s) that you think we should know about, please see appendix.

More: Other sources of essential bacteria: Bloodmeal, liquid Seaweed, Fish emulsion, (Nitron A35, AgriGro, Shure Crop, Superseaweed, Willard Water, to name a few commercially produced), milk, molasses, aged tree bark, Alfalfa meal, herbs. Learn The Bio-Dynamic system. This will help you to increase your basic understanding of how Mother Nature has put everything together and how it works and best of all, how to use it to your benefit.

Tabasco Soap?: Another spray formula to use to control thrips, aphids and most pests: add 1 tablespoon Tabasco sauce, 1 tablespoon natural soap such as Dr. Bronners Peppermint Soap to a quart water spray bottle. This can be sprayed onto the leaves and around base of plants. It will repel and kill thrips, aphids, etc. You can alternatively add 1 tablespoon garden grade DE.

DE: Diatomaceous Earth is a natural dust used to control crawling insects, beetles, ants, aphids, spiders, snails, etc. DE can be used in many ways. Can be added to water, strained and then sprayed. Use 10 tablespoons per gallon filtered water. DE can be made into a paste and then painted around the base of the roses to prevent caterpillars, snails, etc., from climbing to eat the flowers. When using DE, you must be careful to avoid breathing or eye contact, due to its abrasiveness. Garden grade DE is perfectly safe(pool grade DE is dangerous) to use but it is a dust and should be handled with care. Wear a face mask when using and wash with water if it gets in your eye. Do not rub your eyes, remember DE is like thousands of tiny razor blades. Water will wash it away. For more information on DE see index. A good source of DE is Nitron Company.

Plants under the care of the ants are protected by them and are also used as a source of food either for them directly thru the sap or pollen of the plant or indirectly thru the use of aphids and other insects which attack the plant and which in turn are 'milked' by the ants (for their nectar). Changing the behavior of the ants is a very important factor in controlling many pests in the home and garden as well as controlling the ants themselves. This is developing a line of communication between the ants and yourself. Ants are uniquely positioned in the insect kingdom. They are intelligent enough to remember. Ant memory works in a very direct way. They are 'programmed by nature to behave in a certain way. Their actions are controlled by certain factors in their environment. Control these factors and you control the ants. Quiet Control is what you seek. It is not necessary to kill them. For more information on ants see chapter "Dances with Ants".

Getting A Hold: Raise the energy level of the soil; in turn you will have higher energy levels of the rose, vegetable or plant. The higher the energy level of the soil, the healthier the rose, the less stressed and the fewer pests your roses will have attacking them. Ants respond to imbalance and stress.

Give Caesar his Due: Feeding the ants will reduce their activity in other parts of their kingdom(our yard and house). Ants like most creatures except man, follow the path of least resistance. Ants are good at this. If an endless food source has been found, they will use it to the benefit of their colony. Ants will not have to go looking for food as long as they get what they need. The basic idea is to provide for the ants a feeding center, see Ant Cafes™ chapter one. Set up at least one Ant Cafe near by. By providing them a food source, we can begin to retrain them to come for food here and to stop looking for food in the kitchen, or on the roses, etc.. This is a simple system that can become your most effective control method against the ants.

Using Natural Barriers to keep Ants/Aphids off: Tangelfoot works well for this. You can add cayenne pepper to increase its effectiveness. Other barriers to keep ants and aphids off are Tabasco sauce and a natural soap sprayed on base and on leaves. Use 1 tablespoon of ea. in 1 quart water. A good soap to use is Dr. Bronners Peppermint soap that is available at most health food stores. Any natural soap that has a strong fragrance will work, Peppermint is very effective for this purpose. There are many natural fungicides that are used to control Downey Mildew, Rust, Black Spot, Powdery Mildew and other exotic diseases, but by now you should get the idea that the basis for regaining the health of the roses and other plants is to regain the health of the soil. The an important word here is "Exotic". This is very important to understand. "Exotic" diseases can be directly correlated to the lack of "Exotic" trace minerals. Usually it is an absence of, rather then the presence of these exotic minerals that triggers the effect (which is the disease). This is also true for common diseases.

To regain control of any such disease, you must regain the balance of the eco-system that the plant is growing in. Check soil PH levels, Check for excessive salt levels such as boron, chlorine, etc. Check watering habits and equipment. Use a garden filter to filter out the chlorine. Check your composting/fertilizer habits, Switch over slowly to Organics. See more info in Dances with Ants chapter.

Vit C?: Vit C is an effective way of controlling aphids on your roses. Use a liquid vit c and spray several times per week or as needed. Use about 1,000 units per quart of water.

Some Common Diseases of Roses, their Causes and Cures

NAME OF DISEASE: BUTRYTIS BLIGHT

Description of Disease: Brown spots appear on buds' petals may become brown and drop off blooms fail to open, roses with larger petals are most affected. stunts' plants.

Causes: A fungus that likes moisture in air and soil. caused by dead soil, eco-system, chlorinated city water, using high nitrogen fertilizers (urea based), poor health of plants due to improper nutrition over head watering encourages fungi to spread, planting wrong variety /wrong location.

Immediate Solutions: Reduce over head watering, provide soaker or drip apply compost and mulch over. Use a Garden Filter when watering to reduce chlorine in soil. Use Bacterial sprays such as Nitron A-35, Shure Crop, Superseaweed, Agri-Gro etc. Spray a fungus spray such as garlic/baking soda mix (Add 5 drop's garlic oil to 1 quart water, add 2 drops Dr. Bronners Peppermint soap as a wetting agent, 5 tablespoonfuls of baking soda and 1/2 cup apple cider[2]). If you do not have any baking soda handy, you can use 1 cup milk in a quart water instead. Compost tea spray will also encourage beneficial bacteria. see compost tea page. Replace with more resistant variety. Old Fashion heirlooms are best.

Short Term Solutions: Composting/mulching several times per year controlling water thru soaker or drip systems using a garden filter when ever possible

Proper Nutrition: Use no chemical fertilizers, no high nitrogen, use composted animal manure's, use organic fertilizers, use only natural sprays.

Long Term Solutions: Proper composting and mulching, soakers or drips system used best with inline feeder systems, use natural beneficials (like beneficial nematodes), use only organic fertilizers (or make your own), use natural foliars, encourage earthworms, learn companion planting.

NAME OF ROSE DISEASE: BLACK SPOT

Description of Disease: Black spots appear on leaves, expands to larger spots that may turn yellow and whole leaf dies. Entire leaves turns yellow and falls off, mostly affects lower leaves on plants.

Causes: A fungus that likes moisture in air and soil, caused by dead soil, eco-system, chlorinated city water, using high nitrogen fertilizers (urea based), over head watering encourages fungi to spread.

Immediate Solutions: Reduce over head watering, provide soaker or drip, apply compost and mulch over. Use a Garden Filter when watering to reduce chlorine in soil. Use Bacterial sprays such as Nitron A-35, Shure Crop, Superseaweed, Agri-Gro etc. Spray a fungus spray such as garlic/baking soda, add 5 drops garlic oil to 1 quart water, add 2 drops Dr.

---

2 Baking Soda works well against most fungal diseases if used sparingly. Remember that it can affect the soil health as well as the plants health if over used.

A GOOD BIODYNAMIC FORMULA

Here's a very old Biodynamic formula that will help you to control many diseases on your roses, plants etc. Obtain horse, cow, sheep, llama, or rabbit manure. Make sure it's not more then 6 months old. You will need only about 1-5 lbs. The amount depends on the number of roses, etc. you want treated. One cup of this manure mixture will make 3 gallons of a liquid spray. One gallon of this spray will take care of 10 full grown roses, etc. Per every lb. of manure you get add 1 cup rock dust, 1 cup natural clay, 1 cup powdered seaweed . Place this mixture into a large covered clay pot and mix well. You can either bury inground (in a shady area) or place in your basement if you have one. Allow to sit for 1 month. Take 1 cup of this mixture and place into a panty hose. Place into a 5 gallon glass container such as the one mentioned above. You can buy these from your water company. Allow to sit in the sun for at least 1 hour, three hours is best. Remember to use only filtered or solar water.

If you are going to use solarized water, you can add city water to the five gallon container the day before, stirring regularly. Keep what mixture you are not using inside the clay pot. This will be effective for up to 1 month. Make a new batch if longer than that. Spray on the leaves. Works best in early morning or early evening. Spray daily until effective.

Using Molasses

You can use molasses to control many diseases on your roses, etc. Here's a good formula to use. To one gallon filtered or solarized water add 1 cup unsulphured molasses. To this add a dash of soap such as Dr. Bronners Peppermint soap (1 tablespoon will do as it is a concentrate), any natural soap will do, experiment! Spray on leaves. This formula only works on roses that are already being fed properly and have a good alive compost/soil to rely upon. The molasses provide a special bacteria as well as sugars in a form available to the plants.

Bronners Peppermint soap (as wetting agent) and 5 tablespoons baking soda. Compost tea spray will also encourage beneficial bacteria. See compost tea page.

Short Term Solutions: Composting/mulching several times per year. Controlling water thru soaker or drip systems, using a garden filter whenever possible.

Proper Nutrition: Use no chemicals, no high nitrogen fertilizers, use composted animal manure's, use organic fertilizers, use only natural sprays.

Long Term Solutions: Proper composting and mulching is important. Use soakers or drips systems, best with inline feeder systems, use beneficials (like beneficial nematodes), use only organic fertilizers (or make your own), use natural foliars mentioned in this book, encourage earthworms, companion planting, A fine horticultural oil can be used here in the early spring to control spores.

NAME OF ROSE DISEASE: CROWN GALL

Description of Disease: Tumor like growth on canes or roots, most often found on bud unions.

Causes: A Bacteria that enters plant thru cuts or wounds, caused by dead soil, eco-system, chlorinated city water, using high nitrogen fertilizers(urea based), poor health of plants due to improper nutrition, spreads from contaminated tools or soil, planting wrong variety / wrong location.

Immediate Solutions: Remove affected areas of plant, sterilize tools after use (use 1% bleach solution), use soil sulfur to acidify the soil and to provide sulfur to the soil., apply compost and mulch over. Use a Garden Filter when watering to reduce chlorine in soil. Use Bacterial sprays such as Nitron A-35, Shure Crop, Superseaweed, Agri-Gro etc. Spray a fungus spray such as garlic, add 5 drop's garlic oil to 1 quart water, add 2 drops Dr. Bronners Peppermint soap (as wetting agent)and 5 tablespoons baking soda. or any other natural soap. Compost tea spray will also encourage beneficial bacteria. See compost tea page.

Short Term Solutions: Composting/mulching several times per year, controlling water thru soaker or drip systems, using a garden filter whenever possible.

Proper Nutrition: Use no chemicals, no high nitrogen, use animal manures, use only organic fertilizers, use only natural sprays mentioned in this book.

Long Term Solutions: Proper composting and mulching is important, soakers or drips system used best with inline feeder systems, use beneficials (like beneficial nematodes), use only organic fertilizers (buy or make your own), use natural foliars mentioned in this book, encourage earthworms, companion planting.

NAME OF ROSE DISEASE: CANKERS

Description of Disease: Canes turn brown from other then winter dieback, stunts plants, occurs in spring or fall.

Causes: A fungus found in air and soil, caused by dead soil, eco-system, chlorinated city water, using high nitrogen fertilizers (urea based), poor health of plants due to improper nutrition, over head watering encourages fungi to spread, planting wrong variety / wrong location.

Immediate Solutions: Removal of affected parts, sterilize tools between use, reduce over head watering, provide soaker or drip, apply compost and mulch, Use the Garden Transformer (see appendix and index) when watering to reduce chlorine in soil, Use Bacterial sprays such as Nitron A-35, Shure Crop, Superseaweed, Acadie, Agri-Gro etc. Spray a fungus spray such as garlic, add 5 drop's garlic oil to 1 gallon water, add 2 drops Dr. Bronners Peppermint soap (as wetting agent), and 5 tablespoons baking soda. Compost tea spray will also encourage beneficial bacteria. See compost tea in index.

Short Term Solutions: Composting/mulching several times per year, controlling water thru soaker or drip systems, use the garden transformer when ever possible,

Proper Nutrition: Use no chemicals, no high nitrogen fertilizers, use composted animal manure's, feed with an organic fertilizer, use natural sprays mentioned in this book.

Long Term Solutions: Proper composting and mulching is important, use soakers or drips system used best with inline feeder systems, use beneficials (like beneficial nematodes), use only organic fertilizers (or make your own), use natural foliars, encourage earthworms, companion planting.

NAME OF ROSE DISEASE: DIEBACK

Description of Disease: Canes turn brown from tips, plants most affected in colder climates, right after winter protection is removed, not a disease itself but can led to disease attack.

Causes: Caused by early removal of winter protection, new shoot tips are damaged, diseases will attack and enter plant.

Immediate Solutions: Cut off damaged section, spray a natural fungus solution such as garlic, milk, rock dust milk or any mentioned in this book, provide adequate winter protection.

Long Term Solutions: Proper composting and mulching, soakers or drips system used best with inline feeder systems, use beneficials (like beneficial nematodes), use only organic fertilizers (buy or make your own), use natural foliars, encourage earthworms, companion planting.

NAME OF ROSE DISEASE: POWDERY MILDEW

Description of Disease: Leaves fold up, reddish in color, white powdery growth on buds and leaves, blooms fail to open or fall off, stunts plants.

Causes: A fungus which likes moisture in air and soil, caused by dead soil, eco-system, chlorinated city water, using high nitrogen fertilizers (urea based), poor health of plants due to improper nutrition, over head watering encourages fungi to spread, planting wrong variety /wrong location.

Bordeaux Mixture

Bordeaux Mixture works well against most mildews and fungus such as Powdery Mildew and Rust. You can make your Bordeaux mixture by adding three oz. of copper sulfate (blue-stone) to 3 gal. filtered water. Dissolve well. Add 5 oz. of hydrated lime and again mix well. Ready to use at 1 part mixture to 1 part additional water. Can be used at full strength for bad infestations.

Fine Horticultural Oil

Using Sun Sprays Ultra Fine Horticultural oil will smother the rust spores and it will reduce them during the growing season. Fine Horticultural Oil can be used year round without burning the plants. If used during the early spring , it will help to control and reduce many fungi. Available thru Gardens Alive!

Using Baking Soda

Baking soda can be used to control most diseases but should be used as a control tool and not a crutch. Do not get into the habit of using baking soda everytime there is a problem. Fill 3/4 of a cup with water. Add 5 tablespoons of baking soda, stir in well, add 5 tablespoon's vegetable oil or ultra fine oil or fish oil or coconut oil, add 2 tablespoon Dr. Bronners Peppermint Soap. Stir in well to dissolve then add either 2 tablespoons of vinegar or apple cider. Fill cup too full. Stir in well to dissolve, pour either into quart sprayer filled with water (preferably filtered or solarized) or add to hose end sprayer adjust mixture.

Immediate Solutions: Reduce over head watering, provide soaker or drip, apply compost and mulch. Use a Garden Filter when watering to reduce chlorine in soil. Use Bacterial sprays such as Nitron A-35, Shure Crop, Superseaweed, Agri-Gro etc. Spray a fungus spray such as garlic, add 5 drop's garlic oil to 1 quart water, add 2 drops Dr. Bronners Peppermint soap (as wetting agent). Compost tea spray will also encourage beneficial bacteria. See compost tea page. Baking Soda is very effective for this type of problem. Use 5 tablespoon baking soda 1 tablespoon vegetable oil or ultra fine oil or fish oil; 1/2 tablespoon Dr. Bronners Peppermint soap and 1 tablespoon vinegar. Into a cup first add enough water to fill 2/3 of the cup, add the baking soda, vegetable oil, soap, fish oil stir then add the vinegar. Apple cider work well here too. Dissolve together then add to gallon water and pour into sprayer. Spray on leaves of roses.

Short Term Solutions: Composting/mulching several times per year, controlling water thru soaker or drip systems, using a garden filter whenever possible (see garden filter page).

Proper Nutrition: Use no chemicals, no high nitrogen fertilizers, use composted animal manures, use organic fertilizers, natural sprays.

Long Term Solutions: Proper composting and mulching, soakers or drips system used best with inline feeder systems, use beneficials (like beneficial nematodes), use only organic fertilizers (buy or make your own), use natural foliars, encourage earthworms, companion planting. Fine horticultural oil will work here in keeping the spores from spreading. Should be done in the early spring as the earth get warm.

Some Pests of Roses, their Causes and Cures:
Organic Treatment of Leaf Cutter Bees

Description of Damage: Circular holes in leaves, semi-circular holes also. Stem's dieback.

Pest Description: Bees use leaves for nesting, they also bore into stems.

Immediate Solutions: Does not affect healthy plants. Use tobacco sauce/soapy water to control, removal of dead stems, planting society garlic to repel, bury tobacco dust at base of plant.

Organic Treatment of Mossy Gall Wasp

Description of Damage: Green and moss like balls on rose canes

Description of Pest: Wasp egg nest on rose leaves.

Immediate Solutions: Removal of egg nest only if plant is weakened, bury tobacco dust at base of plant to repel, plant society garlic to repel.

Organic Treatment of Rose Aphids

Description of Damage: Leaves are curled up, possible fungus growth. Ants/aphids are present.

Pest Description: Reddish or greenish, usually no wings unless mating season, very small. usually found on buds and undersides of leaves.

Immediate Solutions: Spray 5 drops Tabasco sauce, 5 drops Dr.

Bronners Peppermint soap per gallon filtered water, add Superseaweed and or Nitron A-35, spray a natural fungus spray to control bacterial growths. Try Garlic/soapy water. Compost tea makes a good spray also.

Short Term Solutions: Control ants, see Dances with ant's chapter, control over head watering, avoid high nitrogen fertilizers, apply compost/mulch, plant society garlic to repel, bury tobacco at base of plant. Vit C is also an effective way of controlling aphids on your roses. Use 1,000 units per gallon water.

Long Term Solutions: Use drip or soaker systems, apply compost regularly, control pest vectors such as ants, bury tobacco at base of plant[3]

ORGANIC TREATMENT ROSE BORERS

Description of Damage: Drooping unopened buds, wilted leaves or branches, stunted plant

Pest Description: Small greenish caterpillar like larvae are found in soil

Immediate Solutions: Prune all affected areas, protect cut areas with clay/cayenne pepper, use beneficial nematodes to control larvae, use tobacco sauce/soapy water spray or spray with a tobacco juice/garlic mixture, encourage natural predators.

Long Term Solutions: Encourage healthy soil, attack natural predators, avoid high nitrogen fertilizers, bury tobacco at base of plant, plant society garlic to repel.

ORGANIC TREATMENT : JAPANESE BEETLES

Description of Damage: Leaves are being skeletonized, buds do not open up, stunted plant

Pest Description: Beetle with metallic brown with green head, North Eastern USA only, Grubs feed on lawns until spring then feed on rose leaves in early spring or early summer.

Immediate Solutions: Use pheromone traps, use milky spore disease, use beneficial nematodes to control.

Long Term Solutions: Encourage healthy soil, attack natural predators, avoid high nitrogen fertilizers.

ORGANIC TREATMENT OF RED SPIDER MITES

Description of Damage: Leaves are covered with tiny holes, leaves turn yellow and fall off, undersides of leaves are most affected, tiny webbing present, ants maybe present, possible fungus growth

Pest Description: Microscopic insects, barely visible as reddish specks

Causes: Nutritional deficiencies, high nitrogen combined with over watering produces new green growth that is perfect for Spider mites.

Immediate Solutions: Spray 5 drops Tabasco sauce, 5 drops Dr. Bronners Peppermint soap per gallon filtered water, add Superseaweed and

---

3 Tobacco is absorbed into roses, kills aphids and most other pests. Avoid tobacco that has additives, and other chemicals. Naturally grown is always the best. Grow your own!

BIOLOGICALS

Biologicals are used to control many varieties of rose attacking caterpillars: BT, MVP, Neem(available from Gardens Alive, etc.). Use Tangelfoot (available from most garden centers) and place a line around the base of the plants. It will keep caterpillars (and snails) from climbing. DE made into a paste and painted around base of rose will work also.

SOAP

Using Soap will control spider mites, aphids, ants, thrips and many other rose pests. Use 5 tablespoons of Dr. Bronners Peppermint soap, or any other natural soap, per gallon filtered water. Safer Insecticidal soap is also excellent to use. Proper pruning is very important and can only be learned from experience.

or Nitron A-35, AgriGro, or Roots, or Acadie spray a natural fungus spray to control bacterial growths. Try Garlic/soapy water. Compost tea makes a good spray also.

Short Term Solutions: Control over head watering, avoid high nitrogen fertilizers, apply compost/mulch, plant society garlic to repel, bury tobacco at base of plant.

Long Term Solutions: Use drip or soaker systems, apply compost regularly, control pest vectors such as ants and aphids, bury tobacco at base of plant once per year.

ORGANIC TREATMENT OF ROSE SNAILS

Description of Damage: Leaves are being eaten up, blooms have holes in them, slimy trails present on leaves and buds, snails and or ants are present, possible fungus growth.

Pest Description: Snails or slugs (without shells)

Causes: Nutritional deficiencies, high nitrogen combined with over watering produces new green growth that is perfect for snails, dead over fertilized soils[4].

Immediate Solution: See snail's chapter.

ORGANIC TREATMENT OF ROSE THRIPS

Description of Damage: Leaves are curled up, blooms are discolored with spots, blooms fail to open or are damaged, browning of petals, ants might be present, possible fungus growth.

Pest Description: Very small slender insects, barely visible, usually found on buds and inside curled leaves.

Causes: Nutritional Deficiencies, high nitrogen combined with over watering produces new green growth that is perfect for thrips[5].

Immediate Solutions: Spray 5 drop's tobacco sauce, 5 drops Dr. Bronners Peppermint soap, per quart filtered water, add Superseaweed and or Nitron A-35 or any other natural seaweed or bacterial product, spray a natural fungus spray to control bacterial growths. Try Garlic/soapy water. Compost tea makes a good spray also.

Short Term Solutions: Control ants, see ant's chapter, control over head watering, avoid high nitrogen fertilizers, apply compost/mulch, plant society garlic to repel, bury tobacco at base of plant.

Long Term Solutions: Use drip or soaker systems, apply compost regularly, control pest vectors such as ants, bury tobacco at base of plant.

"Going from chemical to organic can be a traumatic experience to both the rose and the grower. Not only Plants but people go through chemical withdrawals. You must allow at least one year to complete the switch."

---

4 City water contains chemicals to kill off bacteria. Constant use destroys soil organisms, causes imbalance.

5 High nitrogen causes stress!

13

Organic Tree Care

About Trees

Trees are very important to this planet's ecosystem. They are essential to the earth's recycling system. They provide the very air we need to live. They control the water cycles. They provide homes to many different species. They have been called the earth's shock absorber. They assimilate carbon into oxygen, they clean up the air (improve air quality and bring water and minerals up to the surface. Reduce water run off. Urban forestry helps clean up cities and raise the quality of life, promoting interaction between urban dwellers and their environment. Trees also reduce heat and glare.

WHY FEED THEM CHEMICALS?

The answer is simple, trees cannot use chemicals in their system! Trees must be fed naturally and in accord with how the natural system works. Remember trees have been around for a long time.

PROPER TREE CARE

How to feed them then? Chemical fertilizers damage the soils ecosystem, cause stress and lack the essential bacteria and minerals. Trees require certain conditions for proper growth and good health. Proper site selection is very important. It is important to understand what these conditions are. Proper tree selection for the environment it will be in is also important. Before planting amend the soil with good rich compost and lots of humus. Provide regular watering. Provide regular composting.

Mulch as needed: Most trees will mulch themselves so do not rake away their leaf droppings. Do not use any chemical fertilizer or pesticides. Do not top trees! Go Native whenever possible. Hire only skilled workers. Remember, it cost less to keep a healthy tree then to replace it. Save a tree whenever possible. Do not cut it down, plan around and include the tree in your planning!

Trees need water: Make sure that you provide adequate watering on a regular basis. A drip system or soaker will help. Filter the water. Use tree vents for deep watering if possible. Use a garden filter for this. The proper selection of trees is important for this purpose.

Trees love compost: Compost is one of the few things you can feed trees with. Local compost rich with trace minerals, composted animal manure and lots of humus are best. Feed yearly if possible.

Never feed trees high nitrogen: High nitrogen causes stress that is the main cause for pest or disease attack. Trees get their nitrogen from the air, from the compost and from the natural bacteria found in the soil.

Tree vents: A tree vent is a clay pipe that is placed under a tree. Four tree vents per tree equally spaced about 2-4 feet away from the truck (depending on size or age of tree). I use clay drain pipes that are 3 inches wide and 12 inches long, long enough for our purpose. For a very old tree try 24 inches long (Two 12 inch tree vents stacked on top of each other). Run a drip line to water the vents. A 2 gallon per hour drip head in each vent will do. Use only Filtered or Transformer water. Water at least once per month or as needed. Best is a long slow drip over night. Inside the tree vent place compost, rock dust, etc. Also in the tree vents you can place the following mixture if the tree is being attacked by pests or diseases. See info on YBM water transformer in appendix. Do not allow to dry. This is a very important tool in your natural tree care program. Learn to use it right!

The Tree Vent Formula TGM+

Makes enough for four tree vents

1 cup Carbonated water[1]

1 gallon Nitron A-35[2]

1 gallon Agri-Gro

1 quart Superseaweed[3]

1 gallon Liquid Acadie Seaweed[4]

2 lbs Compost

2 lbs Organic Tobacco[5]

2 lbs Rock Dust

½ lb. Garlic

Blend compost, garlic, tobacco, and rock dust, Separate into 2 lbs. Place inside zip lock bag. Allow to sit inside a zip lock bag for one week then add 2 lbs into each tree vent. Add 1 cup ea. of Nitron A-35, Agri-Gro, Acadie to each tree vent, add 10 drops Superseaweed to each tree vent. Add ¼ cup carbonated water into each tree vent. Water well. Do not allow water to overflow. Water slowly to fill. Use water transformer to clean water of any chemicals, very important!. See YBM in appendix. Repeat again in two weeks. Then again in a month. Repeat the TGM+ formula in two months time, then every six months or as needed Do not allow to dry.

Placing the tree vents: Make sure drip system is working. Do not allow to dry. You can use tobacco dust or buy organic tobacco. Tree vents can be used for all trees placing only compost as a food source. This provides for deep watering, encourages root development, provides oxygen to the soil. The tree vents are an essential in pest control since the tobacco will kill any pests that are attacking it. The Garlic will prevent bacterial diseases from spreading. The carbonated water provides trees with a source of carbon dioxide.

NATURAL PEST CONTROL METHODS

Healthy trees will not be effected by insects and diseases as much as sick trees are. That is your first line of defense. However what to do until the tree gets healthier is very important. Plant pests and diseases play an important role in controlling the tree population. Not all tree pests are harmful to the tree - some are beneficial. Learn to identify the symptoms of plant problems, experience will teach you.

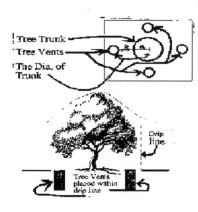

"Tree vents can be used for all trees placing only compost as a food source. This provides for deep watering, encourages root development, provides oxygen to the soil. The tree vents are an essential in pest control since the tobacco will kill any pests that are attacking it."

Here are listings of various possible insects and diseases which attack trees and their Natural Controls, Causes and Associated Nutritional Deficiencies (AND), are listed in order of cause at end of each.

PESTS

Aphids: Aphids attack by sucking tree juices, attract many other insects and cause fungus to spread. Immediate control is soap and water. Add 2 tablespoons of Dr. Bronners Peppermint Soap, or Jungle Rain or any natural soap or insecticidal soap, add 2 tablespoons Tabasco sauce per gallon water. Spray tree. Remember that aphids are attacking the tree because the tree is stressed, is it near a lawn? High nitrogen causes stress. Usually mean lack of water, too much water or a lack of proper nutrition. Provide compost. Other stresses can come from environmental causes. See also ants since by controlling the ants you will reduce aphid populations on the tree also. AND: Need's calcium, iron. See index for info on soap. Tree Vents in use with TGM+ will control problem.

Ants and your trees: Ants and trees have a relationship that dates back for a very long time. We are not here to change this relationship however we are to better ourselves and our environment. It is important to understand that ants are important to the trees. They provide for them the means to reproduce and to grow far away from the parents. But we have also learned that ants respond to stress and imbalance. Our environment is constantly under stress due to the workings of mankind. We can watch the ants and learn from them. The ants will tell you when a trees is under stress and needs help. Ants will climb the trees that are weakened from stress, etc. and bring with them aphids and other creatures that attack the tree. Aphids for example produce a sweet nectar that the ants love. It is also this nectar that brings diseases and other pests. Tree Vents and TGM+ will control problem.

"It is important to understand that ants are important to the trees. They provide for them the means to reproduce and to grow far away from the parents."

To help the tree recover, you must first keep the ants from climbing the tree. This is done in several ways....

1....water can be sprayed from a hose to wash the ants off the tree. Must be repeated to be effective.

2....read the ant control chapter to best determine how to control the ants. Use the ant cafes to do this.

3.....Place a barrier around the trunk of the tree. This will not permit the ant's access to the tree. A good barrier is called Tree Tangelfoot and is available at most nurseries. Before you apply the Tangelfoot, get yourself a roll of dust tape. Place a band of duct tape around the trunk of the tree onto which you place the Tangelfoot. This is to protect the back of the tree from being damaged by the Tangelfoot.

---

1 Pure carbonated water is best avoid chemical additives.

2 Or any other natural earth rebuilding enzyme/seaweed/organic/etc. will do.

3 Or any other chemical free liquid seaweed or powder.

4 Or any other type of liquid seaweed will do.

5 Do the best you can to get organically chemically free tobacco as possible but if you cannot buy a pound from your local smoke shop.

RULES FOR NATURAL TREE CARE

☑ Never Feed trees a chemical fertilizer. Chemicals kills off the bacteria in the soil while at the same time weakening the health of the tree. Chemicals cause a great deal of stress to trees. Trees require regular watering. Use a drip system whenever possible. Avoid watering trunk of tree. Trees love slow deep watering.

☑ Always use Filtered, Well, Rain, Solarized water or Transformed water.

☑ Never use straight city water as it may contain chlorine or some other chemical that can poison the tree. See Garden Water Transformer in appendix.

☑ Allow Time for trees to Heal. The bigger the tree, the longer it will take to heal. People do not realize that. Just remember that we are mother nature now.

☑ Raise the Energy level of the tree. Besides taking a vacation the addition of compost rich in trace minerals and bacteria is the best thing you can do to help raise the tree's energy levels. Trees also love Rock Dust. Also Foliar feeding them will provide for them the fastest source of food and energy therefore it is important that you only feed them nutrition that relieves stress; and not causes it. Understand?[9]

Ants and tree sap: If your tree is leaking sap, this is usually an indication that the tree is under some type of stress and is opened to attack.

Ants are usually the first to attack. Actually what happens is that the ants find out about the tree and herds' aphids up the tree to provide it (the tree) as a food source. Aphids as well as other insects and diseases are attracted to this sick stressed tree. It is important to first find out the source of the stress. Does it need to be fed or watered? Is it being damaged by something? take a good look at it. See if you can understand what it's telling you. Control the ants; keep from climbing the tree. AND See Also Nutritional and Water requirements for trees. Tree Vents and TGM+ will control problem.

Leaf feeding beetles: Can be controlled by using tree vents with 1 lb. TGC+ (Tobacco, Garlic, Compost, Rock Dust). This will take time to be effective, also depends on how well root system works. You can use beetle traps, nicotine sulfate, Sabadilla, Ryania, DE, rotenone is effective also. A fine horticultural spray will kill beetle larvae, or you can use 1 oz. Coconut oil mixed with 1 oz soap, pour into 1 gallon water. If you use beneficial nematodes, or Spined bean beetle, spray only seaweed as using any of the above will also kill the beneficials. AND Key in to proper nutrition. Use compost! Tree Vents and TGM+ will control problem.

For Borers: Use tree vents to control. Use tobacco pushed into holes made by borers. Trees can be sprayed with DE, soap and seaweed. Fill tree vents with TGM+. Water well. AND Provide trace minerals in a compost form. Tree Vents and TGM+ will control problem.

For Leaf feeding Caterpillars: Use BT or MVP add a natural soap and molasses to mixture. AND Lacking Trace minerals. Tree Vents and TGM+ will control problem.

For Leafminers: Use BT or MVP, add soap and molasses. AND: Stress due to lack of trace minerals. Tree Vents and TGM+ will control problem.

For Nematodes:

1......read immediate help with trees page

2......condition of soil

3......addressing nutritional requirements and water requirements

4......using the Tree Vents to control nematodes

5......using Beneficial Nematodes other Clandosan[6] .

For Plant Bugs: Use soap sprays like peppermint soaps or citrus soaps, DE. Tree Vents and TGM+ will control problem.

For Scale: Use soaps. Try 4 tablespoons Dr. Bronner Peppermint Soap[7] per quart water. Don't over do it as too much soap will burn. Scales are usually connected to the presence of ants. Use Tangelfoot[8] to keep ants off the tree. Tree Vents and TGM+ will control problem.

To Control Spider Mites: Use a mixture of soap and DE and Tabasco sauce and a liquid seaweed or Nitron. Tree Vents and TGM+ will control problem.

To Control Tent Forming Caterpillars: Use BT or MVP and soap sprays. Compost tea sprayed regularly will help.. Tree Vents and TGM+ will control problem.

DISEASES TO CONTROL

Cankers: Promote the natural bacteria in the soil. Spray with compost tea and Nitron A-35, Agri-Gro, Acadie, or Superseaweed. Use lots of compost rich in rock dust. Use an acid mulch.

Chlorosis: Tree needs minerals not just iron. Provide rock dust once or twice per year. Use lots of compost and mulch as needed. Use tree vents to feed compost to trees.

Help Fight Tree Decay: Provide tree with rock dust, compost and mulch as needed. Spray with compost tea and Superseaweed or Nitron A-35 or Agri-Gro or Roots.

To Control Fireblight: Provide lots of compost and mulch, spray with compost tea and Nitron or Superseaweed. Sulfur will help here but use a small amount. See chapter on making your foliar spray chapter for how to make compost tea. Fungal diseases are caused by dead soil. This is taken care of by applying compost and mulch yearly. Spray with compost tea, Nitron A-35 or Superseaweed or Agri-Gro.

To Control Gall Formers: Use soap and DE mixed with water. Nutrition is very important. Control moisture.

To Control Viruses: Provide lots of compost and mulch. Spray with compost tea, add a tsp. of sulfur and a tsp. of garlic. Also a good spraying of Rock dust milk and Superseaweed, Nitron or any type of liquid seaweed.

Ants and Viruses: Ants provide diseases the opportunity to establish themselves in your tree. While some Viruses are borne on the winds most are soil borne and carried by insects such as the ants. One way to control viruses is to eliminate a contact points (called insect vectors) such as ants.

Environmental Damages: This is damage due to an environmental reason such as weather variations, chemicals in the air, soil, and water. Other damages include earthquakes, flooding, fires, etc. There is not much that you can do about these except keep the tree as healthy as possible.

Herbicide Injuries: If herbicides were used flush with lots of water. Provide compost and mulch. Spray a seaweed like Superseaweed or use Nitron. A-35, Agri-Gro.

---

6 Use very little and far between.

7 The key thing here is that you can use natural soap to control pests but that you must be certain that it is made up of only natural ingredient's and that you go lightly.

8 You must be careful using this stuff as it could damage the trunk of the tree. I would clean off at end of year and replace(after a time for trunk to heal), you might even think about not using it at all and instead make a paint like mixture that you can apply through paint brush. Try hot barrier's page.

9 The very same rule applies to the human buddy. Beware of foods that you eat that cost you more then what you get from it(i.e. Your body takes more energy to get rid of the stuff then it got from digesting it)! This always leads to poor health (of plants and people).

☑ Know the law of cause and effect. Control the cause and you control the effect. Trees with high energy levels are not attacked by pests.

☑ Know your trees. Plant only trees that will grow in your area. Give your favorite tree a name. Talk to your tree regularly. Plant only other plants that will grow in harmony with each other. Use no chemical fertilizers or you will damage the tree.

☑ Protect trees from dogs. Dog urine is very acidic and will injure the tree's root system. Water, apply compost, mulch. Watch for any signs of problems.

☑ Mulch naturally. The soil will be packed so the water and air may not reach the roots. Apply compost once per year and a layer of mulch. You seldom have to turn over the soil, in fact, trees prefer it if you layer the compost and mulch and do not dig around the base of the tree. Provide plenty of mulch. Most trees naturally mulch themselves.

☑ Protect tree's from people! People cause more damage to trees then any thing else!

☑ Do not poison the soil. Nuff said here.

Natural spraying

Trees love to be sprayed with many natural products. Seaweed is very good for them as is fish emulsion (without urea). Making a milk out of rock dust is very good for them. This provides them with many natural trace minerals, calcium, iron, magnesium etc. Nitron A-35 or Agri-Gro are good to spray them with (for even better results mix the two!). Acadie is a good trace mineral/bacterial spray for trees.

Superseaweed is an excellent overall seaweed concentrate to use. Provide trace minerals and bacteria. Use only 10 drops per gallon. Follow instructions on all labels. See index for more info. There are many natural products on the market these days. Choose carefully what you will use for the trees, ask questions, avoid using chemicals. Follow the Law of the little bit when spraying. If you are un certain about the amount to spray, spray less, more often. When working with trees give them time to respond. Sometimes results will not happen until the following season!

Pesticide Injuries: Flush area with water. Provide compost and mulch. Spray with a liquid seaweed mixture.

Leaf Scorch: Make sure tree is getting plenty of water. Always filter water. Use a drip system to water tree vents.

To Fight Drought Injuries: Use compost and mulch. Use a drip system to provide water. Use tree vents to allow for deep watering and feeding.

For Salt Toxicity: City water contain chlorine, a salt. Use filtered water only on trees. Apply compost and mulch. Avoid fertilizers as they are all salt based.

Winter Injuries: Most trees will recover from winter injuries if strong (hint). Feed well with compost twice per year if possible, spray with rock dust milk or seaweed.

Some Rules for Spraying: A 30 gallon sprayer is best to spray trees. Spray trees either early in the morning or late in the afternoon. Avoid spraying if temperatures are above 90 degrees F. Never spray one particular element, always spray a natural blend of sources. Always deep water trees before spraying. For more information see making your own foliar spray chapter.

Sugar and Trees: Using molasses, corn syrup, etc. as a spray, you are providing trees with a source of energy. Using compost rich in phosphate allows for greater sugar production in trees. Raising the sugar content of trees allows for greater mineral absorption. Use 1 pint per gallon. Add 1 cup of Milk to increase bacterial activity. Calcium is an important source of energy for trees.

H2O2 (Hydrogen Peroxide): Can be used for many different fungal diseases in trees. Use 8 oz per gallon water.

Using Vinegar: Helps to reduce PH level of water, increase energy levels. Use 1 cup per gallon.

Carbonated Water: Increase's carbon dioxide to trees. Use 1 cup gallon.

DE (Garden Grade Diatomaceous Earth): Use ½ lb. per 30 gallon sprayer. Add 1 lb. to bucket, slowly add water to fill. Dissolve DE. Allow to sit then strain into a 30 gallon sprayer. See Using DE in pest control chapter.

Rock Dust: When using rock dust use 1 cup per gallon water. Make the same way you made the DE. Add 1 cup. to bucket. Slowly fill with water and stir to dissolve. Allow to sit a few 1/2 hr and strain into gallon sprayer.

14

Natural Foliar Sprays

Making your own Organic Foliar Sprays

What is Foliar Spraying and how is it important to my property?

The word Foliar means leaves. All plants have leaves of one form or another and when you feed them through these leaves you are foliar spraying them. The idea is to provide plants with important trace minerals, bacteria, enzymes, and elements necessary for a healthy plant to grow. This Chapter will strictly discuss the application of nutritional foliar spraying and not the use of foliar spraying concerning pest controls. That subject is covered in the chapter on Pest controls.

ROCK DUST MILK

Rock dust can be made into a liquid for spraying, see chapter on rock dust, and pest controls and disease control chapters.

ABOUT SEAWEED (KELP)

Seaweed is very high in Potash as well as Magnesium and other more exotic trace minerals such as Boron, Barium, Chromium, Lead, Lithium, Nickel, Rubidium, Silver, Strontium, Tin, Zinc, and even traces of Arsenic, Copper, Cobalt, Molybdenum and Vanadium; all of which are important to proper plant growth as well as proper human growth (over 70 are needed). All the elements that compose this earth can be found in the ocean (if you look hard enough). Seaweed is important for this reason as a trace mineral source in

making compost. It is no wonder that the origin of life was in the ocean.

One of the best foliar sprays comes from Seaweed or Kelp. Seaweed has been used for centuries as a fertilizer to grow mankind's food. The ancients all over the world knew the value of the ocean and her importance to their survival.

To better understand the many ways to use this valuable resource, we must look to our past.

A Trip Into The Past

ISLAND LIVING

Anyone who has ever lived on an island quickly learns the value of the ocean. Being surrounded by water, island living relies on the resources of the ocean. Mankind quickly developed special relationships between themselves and the watery world. Food of course was found in this world. Also found here were the ingredients to help man grow food in the barren island soil. Seaweed was dried and used as a fertilizer, along with anything and everything that came from the ocean.

AMERICAN INDIANS

The American Indian learned the secrets of the water world and passed this knowledge to their children, and children's children. This knowledge was in part the power that came from the ocean that could be harnessed into making plants grow better and produce more. Their method of burying fish for the growing of their corn is well known. This knowledge is still used by many people.

PERU

The barren land would not give to them unless they gave to her. Thus the relationship between the peoples of Peru and the oceans would become a very important one. Peru has a very ancient knowledge 'pool' from which to choose. It is from this 'pool' given to them from ancient times, that they learned how to use the soil. By giving it the richness of the ocean, the barren land would give untold food back in return.

CENTRAL AMERICA

The Mayan, and their precursor's the Olmec, flourished for a millennium along Mexico's Gulf coast. The levels to which they brought agriculture and the special relationships they developed with the ocean and the lands bear a closer look. To understand the reasons why this culture flourished completely intertwined with the ocean and the earth around them, we must look at their relations with those around them.

Even though they were not travelers over water, they traded with those that were and thus were able to extend their ability to control the Earth's production of valuable food. Their land was well suited for food production and they quickly developed special food producing areas that were used completely for this purpose. They developed special "Gardeners" who did nothing but work on these fields. They knew the importance of the ocean's nutritional values. They dried their own seaweed, as well as various ocean creatures such as fish, squid, etc., which they often traded with the fishermen of nearby Honduras, Nicaragua, El Salvador, and even little Belize. It was during the Classic Mayan cultural period (A.D. 300 to 900) that the Mayans developed the art of agriculture to a high degree.

EUROPE

Europeans were one of the earliest cultures to formally recognize the importance of the ocean to their agricultural practices. This was due largely to the masses of people that needed to feed themselves and to their ability to learn from others. They used the ocean's largesse to increase their production along with using animal manure (of which they had a lot), which formed the basis for their agricultural production up until the chemical revolution in the early 1900's. These folks also learned fast from other cultures, as they were able seaman and traveled around the world, and would see many wonderful things and this knowledge would return to their own lands, where it quickly would spread like fire upon the land.

CHINA

The mother land has known the secrets of the ocean since time itself began. China is old and so is her knowledge of the ocean. All the countries of this area knew the ocean- India, Japan; all have learned to rely on the ocean and her bounty. The ocean is very rich in trace minerals, enzymes, and bacteria as well as small amounts of nitrogen, phosphorus and potassium. This special mixture, when applied to plants/soil, can release locked up minerals; improve the cation-exchange in the soil, and has chelation abilities.

"All plants have leaves of one form or another and when you feed them through these leaves you are foliar spraying them. The idea is to provide plants with important trace minerals, bacteria, enzymes, and elements necessary for a healthy plant to grow."

NATURAL SEAWEED PRODUCTS AVAILABLE FROM **IGA** COMMERCIAL MEMBERS:

Kelp Extract: This is a special kelp extract designed for foliar spraying, is made from cold processed Norwegian Ascophythllum Seaweed. Cold processing preserves higher levels of minerals and growth hormones than is found in other kelp extracts. This kelp extract is one of the most potent extracts available on the market today! A definite on your making your own Superseaweed list!! Available from **Acadie, Arbico, Gardeners Supply, Peaceful Valley, Nitron** and from many mail order catalogs.

Maxicrop: Maxicrop is an excellent kelp extract containing over 70 trace minerals, growth hormones, cytokinins, auxins, vitamins, and enzymes. World famous as a foliar feeder and plant stimulant. Contains 1% N, 0%P, and 3%K plus elements from the ocean. Comes as a water soluble powder that mixes easily with water. Use ½ teaspoon per gallon or also available as a liquid concentrate. A must in making your own Superseaweed! Available from **Arbico, Peaceful Valley** and most mail order catalogs.

Sea Crop/Sea Mix: One of the best liquid seaweed's around. A must for making your own Superseaweed! (Seamix is fish emulsion and liquid kelp, highly recommended!!) Available from **Nitron Industries (one of the few around without urea), Peaceful Valley, Arbico.**

OTHER PRODUCTS FROM THE OCEAN

Acadie Fish Meal: Acadie is a good source of all major trace minerals and major elements. Available from **Acadie Seaweed Company.**

Nitron Fish Emulsion: A high Quality Fish emulsion. 5-0-0 A great Foliar feeder, no urea! Available from **Nitron Industries**. No Urea! One of the best!

S Quanto's Secret 2-4-2: Fish Emulsion made from white fish. Available from **Gardener's Supply.**

SuperSeaweed™: Developed by The Invisible Gardener. The ocean has always been a source of nourishment for the plant and animal kingdoms. The Oceans provide us with all the trace minerals, enzymes and bacteria necessary for healthy human growth. Plants need to have the same nutrients we do. Ingredients: SUPERSEAWEED is a blend of over five liquid seaweed's from around the world, selected especially for their purity (no heavy metals, toxin's, etc.). NOTHING TO POLLUTE OURSELVES OR THE ENVIRONMENT EXISTS IN THIS PRODUCT. SUPERSEAWEED is a wetting agent. It makes water "wetter", allowing plants to absorb more nutrients.

Because your plants will be using their available fertilizer to capacity, you will be reducing wasted, or unabsorbed fertilizer. Therefore, only ½ to 1/3 of your present dose of organic fertilizer will be necessary. In itself, SUPERSEAWEED IS NOT A FERTILIZER. SUPERSEAWEED is a special "biological activator"... by making nutrients more available to your plants, they will become stronger and healthier, reducing plant stress, pest problems and fertilizer costs. Plants last longer and look better.

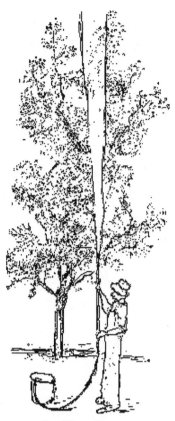

Hand Pump Sprayer

"Choose carefully what you will use for the trees, ask questions, avoid using chemicals. Follow the Law of the little bit when spraying. If you are un certain about the amount to spray, spray less, more often. When working with trees give them time to respond. Sometimes results will not happen until the following season!"

SuperSeaweed™
*Developed by The
Invisible Gardener.
"The ocean has
always been a source
of nourishment for
the plant and animal
kingdoms. Use*
SUPERSEAWEED
*everywhere; flower
and vegetable gar-
dens, houseplants,
fruit trees, hydropon-
ics systems; every-
where!"*

*Nitron A-35: En-
zymes is what Nitron
is all about. Enzymes
is what life is all
about, too. Without
enzymes there would
be no life on this
planet.*

SUPERSEAWEED compliments both chemical and organic fertilizers. I, of course recommend using organic fertilizers whenever possible. Use SUPERSEAWEED everywhere; flower and vegetable gardens, houseplants, fruit trees, hydroponics systems; everywhere! INSTRUCTIONS FOR USE: Do not shake! Use 1 capful per gallon water for all plants. Use once per week. SEEDS: 1 drop per/gal; soak over-night to promote growth. COMPOST STARTER: 1 capful per quart water per 1/2 cubic yard; to help make rich humus. GREENHOUSE: 2 capfuls per gal. twice per week; use with inline feeder to spray leaves. VEGETABLES: 1 capful per gallon, once per week; use only ½ usual organic fertilizer. LAWNS: 5 capfuls/gal. Excellent for orchids, roses! (5 capfuls per gallon.) Available from: The Invisible Gardener. See application and order page 177.

MAKING YOUR OWN SUPERSEAWEED

Here is a formula that will help you to make your own "Superseaweed". This is not the same SS that I make and sell but it will give you great results:

2 part Nitron A-35

2 part Roots

2 part Maxicrop (made from powder)

2 part Acadie Seaweed

2 part Sea Crop

2 part Agri-Gro

1 part Kelp Meal

1 part fish emulsion (optional)

1 part Willard Water

1 drop wetting agent (Dr. Bronner's)

AN EXAMPLE FORMULA

2 quarts Nitron A-35 Formula

2 quarts Roots

2 quarts Maxicrop (made from powder)

2 quarts Acadie

2 quarts Sea Crop

2 quarts Agri-Gro

1 quart Kelp liquid concentrate.

1 quart fish emulsion

1 quart Willard Water

 5 drops Dr. Bronners Soap (wetting agent)

10 quart's total of your very own "Superseaweed"!!

Add 20 drops per gallon for foliar sprayings, as well as watering of houseplants, etc. Use once per week. For sick plants, trees, use 1 cup per gallon, spray once per week, until plant recovers, then once per month. For lawns use 5 cups per gallon for sprayings. The idea in making your own Superseaweed is that in blending the different liquid seaweeds, you would end up with greater amounts of different trace minerals, bacteria, etc. So when following the above mixtures, please feel free to try the various different types of seaweed's on the market and see which works best for you and your particular situation. Some seaweed's are better for fruit trees then others, as some liquid seaweed's are better at warding off pests then others.

Making your own trace mineral spray (Formula #1)

1 Oz Epson Salts

1 cup Greensand

1 cup Rock Dust

1 cup Powdered Acadie Seaweed

1 cup instant Coffee

1 cup Organic Alfalfa Meal

1 cup Gypsum

1 cup Brown Sugar or Molasses

1 cup Apple Cider or Vinegar

Mix powders together, place a cup of the mixture into panty hose and allow to soak in 1 gallon solarized or filtered water, allow to soak over night. See Water Transformer in appendix. Squeeze ball into water, strain into gallon container. Label Trace mineral spray. Use right away. The strength depends on what you are spraying and for what purpose but I would start off with an initial 50% tea 50% water mixture into sprayer. Spray onto leaves in early AM if possible, or early afternoon. Apply at the rate of once per week as a foliar feeder. Best applied during winter months (if possible) otherwise use early spring and then once per month. OK to use fish emulsion or to add more liquid seaweed (following manufacturers' instructions) like Acadie, or to add Nitron A-35, Agri-Gro or, during summer months only, as you do not want to encourage new growth during winter months. The apple cider helps to acidify water.

Making Your Own Enzyme / Bacterial Sprays

Now that you got the minerals you will need the enzymes to eat it!

Nitron A-35: Enzymes is what Nitron is all about. Enzymes is what life is all about, too. Without enzymes there would be no life on this planet. Enzymes are positively charged ions. They perform many tasks. They are important in the detoxification of the soil; soil that is dead from the chemicals poured on them. They help to soften the soil and allow deeper root systems. Enzymes allow water to penetrate deeper into the soil. Enzymes are important to plants in that they help to release minerals, nutrients that are locked up in the soil. This is a must to include when making your own Superseaweed.

Agri-Gro: Invented by Dr. Joe C.Spruill. Ph.D. Biochemistry. A biological complex derived from natural compounds, processed through

---

Developing a Spraying Program

Here is a spraying schedule that you will learn to adapt to your own use. Follow it and learn from it.

Winter: Winter months spray trace minerals on bark of trees, onto leaves of plants. Spray once per month. Add very little bacterias to sprayings as they will not be useful in the winter months, even in warm climates. Use formula #1 for this. Note : this is for southern areas with warm winter. You do not spray during snowing months.

Spring: Introduce bacteria as early as possible. Apply compost tea in sprayings. Use **Nitron A-35**, **Agri-Gro**, **Roots** or **Acadie** to boost bacteria counts along with the compost tea. Add 1 lb. rock dust/5 lbs Maxi Crop to 5 gallon's water, or use your own Superseaweed. Use a gallon of this mixture per 5 gallons of water. This is the spraying mixture you will use. If you cannot make your own Superseaweed, you can obtain the real SuperSeaweed from The Invisible Gardener. See Appendix on Superseaweed.

Continued

Continued

Summer: To 5 gallons water add the following:

½ cup rock dust

2 cups seaweed mixture (see making your own SuperSeaweed) or use Acadie

compost tea (placed into panty hose tea bag)

1 cup fish emulsion (optional)

Allow to sit over night, strain into sprayer.

Fall: To 5 gallon's water add:

1 cup Fish Emulsion

1 cup Agri-Grow

1 cup Nitron A-35

1 cup Acadie

1 cup of your own liquid seaweed mixture

1 Tea bag of compost

1 cup. Rock dust

Allow to sit over night, strain, spray!

Spraying Tips and Hints

Don't over do it! A little bit goes a long way. Learn what you can mix and cannot mix; start your own mix, no mix list. Remember what works. Keep notes. Tell others about your successes!.

extraction completed by fermentation. A plant and bacterial stimulant, contains stabilizing nitrogen-fixing bacteria, trace minerals, and humic acid. Designed to improve chemical and fertilizer efficiency, improves natural sugar levels of plants, improves tilth and water capacity, designed to reduce cost of production, reduce insect and disease problems and reduce soil compaction and also to reduce nematodes and salt buildup in soil. A natural Solution containing living organisms, enzymes, azotobacter, bacillus, clostridium, humic acid and trace minerals. AGRI-GRO stimulates plant growth, promotes health, maintains the beauty of the plant, loosens the soil, is totally "chemical" free and therefore is safe around humans and pets.

Compost Tea: This is one of the most important bacterial sprays you can ever learn to make!! Compost is rich in just the right types of bacteria necessary to proper plant growth. Place 1 cup compost into a panty hose and place into gallon of solarized or filtered or spring or even rain water[1]. Place in sun for a few days depending on the smell. You want it to have a slight odor[2]. Spray in early am or late afternoon. This is an excellent cure for many bacteriological problems such as fungus, etc. Plus help's control many pests. I would make sure that your compost is alive and healthy and not old and lifeless. Make sure you add rock dust to your compost or add 1/2 cup to the panty hose along with the compost to increase the calcium and minerals needed for the bacteria to work properly. Do not forget the molasses! See index info on how to make.

Humic Acid: Comes from Leonardite Ore, millions of years old. Rich in humic acid that helps to break down Organic matter. Available from: Nitron, Arbico and many mail order companies.

A Special Wetting Agent: Here's a little secret. Use Dr. Bronner's Soaps or Amways LOC as a wetting agent for your sprays. You only need to add 5 drops per gallon. Works great!!!.

Oxygen: Plants need oxygen too. Try this.....Go to the drug store and buy the 2% Hydrogen Peroxide[3] ..Add 1 oz per gallon to your spraying mixture. Your plants will love it. This oxygen will allow greater absorption of nutrients into their system.

Apple cider is a good addition to your bacterial spraying kit! Apple cider or apple vinegar will help to increase the acidity of the water and allow bacterial sprays to work better. Only works where your water is alkaline.

---

1 If you must use the city water, I would allow it to sit in the sun for a few days, stirring every once in a while. This will remove most chlorine etc. which are added to kill bacteria!

2 Actually you want the smell to be as strong as you can handle it! The stronger the better!

3 Food Grade or Farm Grade is better.

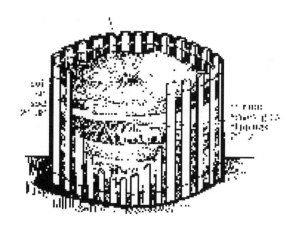

15

Making Super Compost

Compost it!

The wheel of life just keeps on turning. From Death comes new Life. The art of making compost is really a science, nature's science. As a healing ointment for the soil, Compost is the result of nature's dedication to balance. Through watching and learning about the way nature makes compost, we can begin to understand our very own lives and why things effect us the way they do. From Sir Howard to today, the art of composting has taken great steps towards a better understanding of the real process that goes on inside a compost pile.

PURPOSE OF COMPOSTING

There are two major reasons for making compost: 1) To decompose materials down to a more workable and acceptable form. 2) To combine materials in a favorable manner so as to increase nutritional levels as well as increase beneficial bacteria activity, which when added to the soil will increase soil structure.

HOW COMPOSTING WORKS

Bacterial action in a compost pile is the basis for proper composting. It breaks down the various organic matter, converting it into compost while giving off heat and steam in the process. It also makes available trace minerals to the plants. Always provide plenty of oxygen (by stirring pile, aerate it) and plenty of natural nitrogen and trace minerals. Water is very important; too much water will drown and not enough will kill off the bacteria. There are many different types of bacteria present in a compost heap depending on the time from the start of the pile. Different types of bacterial composition will occur throughout the composting process. The time of year also determines the rate at which the composting process occurs.

THE PLANT KINGDOM

From the plant kingdom you get humus. Some sources from the plant kingdom: hay, leaves, grass clippings, apple skins (ash), banana skins (ash), coffee grounds, cottonseed meal, peanut shells, pine needles.

THE ANIMAL KINGDOM

Obtain manure from any or all if you can of: Cattle, chickens, horses, sheep, llamas, goats, rabbits, and ducks. Best is aged at least 4 months. Trace minerals also come from various sources such as: Oyster shells, coffee wastes, silica sand, bay crab meal, kelp meal, (hair is a good source of trace mineral but should be composted well before using) (hoof and horn meal, bonemeal, and blood meal are all optional). Animal manures are also rich in various trace minerals (always vary your manure source).

Heat is important in compost making. When making bins or piles, the bigger the pile the more heat it will be capable of producing. Too small a pile, and too much heat is lost. Bacteria love heat and work best within a range of 140-180 F. Too big a pile will compress the material too tightly and make decomposition a slow process. Also too much heat will kill off bacteria. 4 to 6 ft tall is the proper height for a compost pile. This pile will shrink as composting takes place. This method can also be done during the winter time as the pile is big enough to generate its own heat. You monitor the temp to make sure the pile heats up enough.

About Nitrogen. A proper Carbon-Nitrogen ratio is important for proper composting. Nitrogen is needed to help heat up the compost. Carbon is needed as the fuel. Adding natural sources of Nitrogen (in the form of alfalfa meal, manure or any organically nitrogen rich material) will increase its energy level. Allowing for great activity of the bacteria present. Moisture is important in the composting process, but it must not be soggy. I would not concern myself with this ratio; just work with what you have, keeping in mind the balance required to make the compost pile work right. The ratio will come out just right if you apply a little bit from each kingdom as mentioned above.

SOURCE	%Nitrogen	%Phosphate	%Potash
Ocean			
Shrimp parts (dried)	7.8	10.0	3.4
Dried ground fish (dried)	8	4.0	2.0
Lobster waste (dried)	2.9	1-5	2-3
Crabmeal (dried)	10.0	2-4	1-3
Crab Meal	2.0	1.0	4.0
Fish Meal	5-7	5-7	2-4
Seaweed Meal	.05-2	1-5	5-8
Liquid Seaweed	1-2	1	2-5
Animal			
Dried blood	10-14	1-5	1-4
Feathers	10-15	2-5	05-2
Feather Meal	10-13	2-5	0.5
Eggshells	1.0	.5	.05
Wool wastes	3-6	1-3	05-2
Leather wastes	2-4	1-2	1-4
Bonemeal	0-2	5-12	0-2
Animal Urine	10-15	1-4	1-2
Llama	3.5	2.4	0.8
Rabbit	2.4	1.4	0.5
Chicken	1.1	0.8	0.9
Sheep	0.7	0.3	0.9
Steer	0.7	0.3	0.4
Horse	0.7	0.3	0.6
Duck	0.6	1.4	0.5
Cow	0.6	0.2	0.5
Goat	1.2	1.3	0.7
EarthWorm Castings	1.0	6.0	1.2
Bat Guano	2.0	8.0	0.5
Plant			
Alfalfa Meal	3-5	1-2	1-2
Cottonseed Meal	6-7	2-4	2-8
Cocoa Bean Hulls	3-5	2-4	2-4
Coffee Grinds	4.0	2.2	5-8
Coffee Hulls	2.0	4.0	4.0

SOURCE	%Nitrogen	%Phosphate	%Potash
The Mineral Kingdom			
Granite dust	NA	2.0	3.0-5.5
Greensand	NA	2.0	7.0
Gypsum	NA	1.0	6.0
Rock Dust	NA	2.0	3-7 *

*see rock dust chapter. Greensand, oyster shell, rock, phosphate, limestone, granite dust, and dolomite all provide many sources of trace minerals needed by the plants for healthy living. Added to your compost this will increase mineral etc.. Go to your local fish market, there you can get shells that you can dry and smash into a powder using a hammer and an old burlap bag.

A WORD ABOUT HUMAN HAIR

Human hair contains nitrogen and other trace minerals. Because of the amounts of chemicals used on hair these days at the beauty parlor the hair will not properly compost. Therefore use hair from a men's salon instead of a woman's, as most men do not use these chemicals. Hair also takes too long to compost within 60 days and therefore should be used in the sheet method described later.

TO BIN OR NOT TO BIN?

When you begin to take into consideration what you have available to work, you will then have to decide whether to go the bin route or the pile route. IN short if you need a lot of compost then the piles are easier and faster than using bins. Bins on the other hand work fine for a person with a small garden/home. Three bins min. are suggested. They should be at least 3 ft by 5 ft by 4 ft high and made of wood. There are many books available on making compost bins.

IS IT ROTTING OR NOT?

If it does not smell bad and looks good and dark like rich earth then it's done!

Methods of Composting
30-60 DAY METHOD

1st Day: First in center place a large pile of animal manure such as horse or cow or rabbit or Llama pellets. The size should be about 2 feet high. Water lightly. Place the following layers: Layer in plant waste (hay, grass clippings, leaves etc.). Use your own grass clippings, and leaves to make sure toxins are low (or ask if using someone's else's grass clippings). Sprinkle a thin layer of either alfalfa meal, nature meal or composted manure (do not use cat or dog manure) or any nitrogen rich material. Spray this with a liquid seaweed mixture. You can use Superseaweed™ for this as it makes a good compost starter. Add 1 cup Nitron A-35, or Agrigro or any other natural product. Add a 2 in layer of rock dust covering with a 2 in layer of top soil. Water to moisten. Superseaweed™ is available at: The Invisible Gardener Inc. Other Compost Starters are available at the follow-

ABOUT COMPOST TEA

Compost tea is not a special tea that people drink, but rather a liquid that is made from compost. The best way to make compost tea is to place a cup of compost into a panty hose and tie into a ball. This will act as a tea bag and when hung in water it will spread into the water. The thing to do is to tie the compost ball onto a stick and drop into a gallon(or larger) container of water. Glass bottles make the best compost tea.

Colored glass like wine or beer will also make good compost tea containers as they absorb the sun's energy. The container should be placed in the sun. The time depends on what you are using it for.

The information below tells you the length of time to make compost tea and the purpose

1 hr...............Normal compost tea spray, good for bacterial sprays, mineral sprays. will help to prevent diseases on most plants and trees. Should have no odor.

4 hrs..............A stronger mixture, excellent for controlling most stubborn bacterial diseases. Works great on roses, most vegetables

Continued...

Compost Tea Cont.

and trees.

24 hrs.............A powerful mixture for controlling rust diseases, blight diseases, also repels most insects. 48 hrs............This is good for very bad fungus related diseases. Should not smell bad, but should be strong smelling.

You can increase effectiveness of the above mixture by adding one or more of the following combinations (depending on use): **Soap** will increase effectiveness as a wetting agent, use **Dr. Bronners Soaps** for this purpose. **Molasses** can be sprayed along with the compost tea. The trick is to add 1 cup of the molasses per gallon water after it has sit for a few days. **Vinegar** can be sprayed to increase the energy of the plant reducing diseases and pests. Use 1/2 cup in a gallon compost tea or no more than 5 cups in 50 gallons compost tea. Important note: Use filtered water instead of city water (which has chlorine that kills bacteria). If you do not have a filter use a water transformer (see appendix), use spring water or solarized water. Solarized water can be made by placing water into a large glass container (allow the most sun light to contact the liquid) but a large bucket will do. Allow to sit for a day or two, turning every few hours.

ing outlets: A-35 Nitron Industries, Inc.; they make various compost starters and compost bins. Nitron A-35, Agri-Gro, Arbico. Aerators and many other organic products, Peaceful Valley Farm Supply; another good source of compost starters, thermometers. and various organic materials, Gardens Alive!; they carry many composting products, Mellingers, Inc.; they carry a big line of compost starters, organic materials etc. See Also Arbico for composting supplies. When you write to these folks, tell them the Invisible Gardener sent you!

2nd Day: 1 in layer plant waste or kitchen wastes, 1 in layer alfalfa meal, water to moisten, 1 in layer kelpmeal or dried seaweed, 2 in layer manure (rabbit, goat, Llama pellets), 2 in layer plant waste (leaves, pine needles), 1 in layer minerals (rock dust or granite, gypsum, or greensand, oyster shell), 2 in layer manure cover, 1 in layer top soil, water to moisten, 1 in layer rock dust (a good fertilizer and anti fly dust), 1 in layer kelpmeal, Spray with a liquid seaweed such as Superseaweed™. Cover with a clear plastic or tarp.

3rd day: Allow to sit.

4th day: Allow to sit, take temp (with soil thermometer), leave pile covered, add a ½ in layer of rock dust covering 1 in layer top soil, water to moisten.

5th Day: Allow to sit.

6th day: Take temp, temp should be rising every day till reaches 160 F. The temp may go higher but should stop no higher than 180 degrees F. If the temp does cont. to rise add water to cool down.

7th day: Allow to sit, check temp.

8th day: Uncover pile and allow to sit open for a day, check temp, check for odors (should smell of ammonia).

9th Day: Turn pile over, spraying with the liquid seaweed as you go, turning the pile: This is done by one of several methods...the easiest being using a front end loader to turn the pile over; or use the shovel technique. The main idea is to turn the pile over in such a way that you are turning the layers together except for the center of the pile. **The easiest way to do this is the volcano method (see box).**

Day 10-15: Allow to sit, Check temp every two days, if not hot enough add nitrogen rich materials such as Bloodmeal, if too hot add water to cool down then add more humus or soil to increase nitrogen: carbon ratio.

Day 16: Uncover pile, check temp, turn pile as described before . add 1 lb. alfalfa meal, 1 lb. cottonseed meal use only organic, 5 lbs rock dust, 5 lbs kelpmeal, ½ lb. Epson salts, 1 lb. greensand. Stir together in bucket and sprinkle a thin layer as you turn over pile. Then spray liquid seaweed. The pile should be turned over completely, making sure pile is not too wet (soaky) and not, dry but moist. Cover pile with plastic if possible.

Day 17-21: Allow to sit, checking temp every three days. Keep notes on the temp and compare. Is it still rising or is it falling? The temp should continue to rise until it reaches 160 or higher. Then the temp will level off.

At the point where the temp starts to drop that is the time to turn the pile over again. This usually happens around the 20-30 day, depending on the weather.

Day 22nd: Uncover pile and turn over while spraying with the liquid seaweed. After this the plastic cover is not needed.

Day 23-30: Allow to sit watering lightly to keep moist. Make sure to spray with a liquid seaweed mixture.

Day 31: Turn pile over lightly, (aerate the pile by turning over the top layers).

Day 32-45: Allow to sit checking the temp every 4 days.

Day 46: Turn pile over while spraying with liquid seaweed mixture

Day 47-60: Allow pile to sit, water once every three days. Check temp every time before watering.

Your compost is ready!!

WORKING THE BIN METHOD:

THE 14 DAY METHOD

Day 1: The bottom layer is shredded manure, The next layer is of plant waste shredded.

The next layer is of rock dust. The next layer is of Bloodmeal and bonemeal mixed (optional, kitchen food wastes OK here instead). The next layer is of old shredded manure such as horse. The next layer is of top soil. The next layer is of kelp meal or seaweed meal. The next layer is of rock dust. Then add of either rabbit, or goat (shredded) or earthworm castings. Then add either grass clippings, leaves or both mixed and shredded. Then add a layer either of old horse or old sheep or old cow manure. Then add a layer of Dia-earth (for fly control). Top it off with a 4 in layer of top soil. In between each layer you should add the following: mix together equal parts: Cottonseed meal, Alfalfa meal Kelpmeal, Epsom Salts, Rock dust or Greensand or Rock Phosphate, Compost.

The idea is to lightly sprinkle a thin layer of this formula in between each layer, while also spraying (lightly) a liquid seaweed mixture such as a Superseaweed™ and Nitron™ and Shure Crop mixture (1 cup each added to 1 gallon water (20 drops of SuperSeaweed). Use filtered water. Your compost bin should have a lid and be made of wood to allow it to breathe.

Day 2: Allow to sit.

Day 3: Check the temp. Write this down in your log book.

Day 4: Turn contents of bin #1 into bin #2 (you should have three bins for this purpose), turning over well and adding a sprinkling of the formula as you go and also spraying with the liquid seaweed. Start the process over again in bin #1.

Day 5: Check the temp in both bins. Note this in your compost log, Turn over bin #2, making sure everything is turned over well while adding a thin sprinkling of the formula. Check to make sure both bins are not too wet or too dry. Add water as needed by spraying (this water is the same liquid seaweed I've been talking about).

THE VOLCANO METHOD

This is when you dig into the center of the pile making it look like a volcano has just erupted, then you throw everything back into the center of the pile mixing it up and watering lightly as you go. Cover the pile with clear plastic.

The clear plastic has several functions:

1: It solarizes the compost destroying harmful bacteria.

2: Solarization breaks down various inorganic compounds and allows for their disposal.

3: The clear plastic is a good control against flies laying eggs in your compost pile. The plastic also protects against rain.

*"If it does not smell
bad and looks good
and dark like rich
earth then it's done!"*

*"Through watching
and learning about
the way nature makes
compost, we can
begin to understand
our very own lives
and why things effect
us the way they do."*

Day 6: Allow both bins to sit

Day 7: This is your first week! Isn't compost making fun!!! Today you turn over both compost bins, checking for moisture and heat.

Day 8: Then the next day dump the contents of bin #2 into bin #3, turn bin #1 into bin #2 and start all over again with bin #1.

Day 9: Allow everyone to sit and rest.

Day 10: Turn over bin #3 (sprinkle with formula mixture). Turn over bin #3 (sprinkle with formula mixture). Check all bins for water, temp.

Day 11: Allow everyone to rest today, checking for too much water etc.

Day 12: Check on bin #3, how does the compost look? are there any parts that don't look decomposed? remove that and place in bin #1. Turn over bin #3 ,water as needed with liquid seaweed and add a sprinkling of the formula. Check bin #2 , how are we doing here? Turn over and sprinkle formula and add water (liquid seaweed) as needed.

Day 13: Two more days to go...Allow all bins to sit and rest today

Day 14: Well here it is the day you've been working for. Check bin #3 and tadum!!! Rich Compost. Never add raw waste into bins #2, #3. These bins are for finishing the composting process. Bin #1 is where you may also add kitchen waste such as rice leftovers, salad leftovers etc. Never add meat or cooking oils as this will stop the composting process and you may even have to dispose of the compost. You may add egg shells, coffee grinds, to the first bin only. Rock Dust is excellent for helping the composting process.

16
Rock Dust

The Elixir of the Earth

To grow a greener lawn, have healthier trees, and cultivate bigger vegetables, the soil needs to be enriched. Before reaching for that bag of nitrates or other chemical fertilizers, the conscious homeowner or farmer should stop for a moment to consider what needs to be put back in the soil to enhance its life-giving properties. Like the magnificent form of the human body, the earth has the wondrous capability of healing itself. When forested areas of the world use up the nutrients in the soil, the earth has a built-in remineralization system that can be learned from and applied right in the back yard. The process is known as remineralization through rock dust application. John Hammaker, a research scientist in Massachusetts, postulates that each ice age in the history of the earth regenerated its topsoil. When the planet's forests deplete the soils of nutrients, they begin to die off, releasing their stored carbon into the atmosphere. All this carbon builds up in the atmosphere, creating a "greenhouse effect" and causing a rise in the earth's temperatures.

This heat is most concentrated around the equator where the sunlight is greatest. The higher temperatures cause evaporation, and the moisture rises, creating a vacuum underneath which pulls in cooler air from the polar regions. As the air is pulled in from the North and the South, another vacuum ensues, pulling the warm moist air towards the poles. During this time there is a lot of strong activity like typhoons and hurricanes that occurs in the earth's subtropical regions.

When the moist warm air from the tropics arrives in the polar regions, it hits the cooler temperatures and condenses as snow. The snow builds up, and the weight causes it to pack into the ice and push southward from the north pole and northwards from Antarctica. At this time there is an increase in earthquakes and volcanoes caused by the extra weight on the continents. The glaciers descend, grinding up all rocks and mountains in their path, remineralizing the soil. When the forests again take root, they absorb the carbon from the atmosphere through photosynthesis, and the ice age diminishes.

Hammaker has extended this hypothesis to explain what is currently occurring on earth right now. According to Hammaker, earth's inhabitants have accelerated the onset of the next ice age through the burning of fossil fuels and the deforestation of large forested areas like the Amazon. Hammaker says the only way to stop the glaciers from knocking down our back doors within this generation or next, is to reduce our dependence on fossil fuels, reforest the cities and the country side. By replanting and remineralizing the earth through rock dust applications. While Hammaker advocates going out and remineralizing all forests and fields, the average homeowner usually does not have the time to take on such a large task. However, simply by taking care of one's yard and garden through natural means, the accumulative effects will yield significant results on a global scale.

Some notes about
<u>*KELZYME*</u>*:*

⌘   *Make sure you*
add Kelzyme as it has
the highest source of
calcium!

⌘   *Kelzyme is 32%*
Calcium! Very
important in
reducing stress,
reducing pests!

⌘   *Use Kelzyme on*
the lawn.

⌘   *Kelzyme is very*
good for the garden.

⌘   *Kelzyme is good*
for your compost.

ABOUT ROCK DUST

The amount of minerals and the quality rock used to produce rock dust depends on the location of the for quarry, and the mining process. Many companies sell rock dust, but it is best to inquire about what minerals it contains and if there are any chemicals added. If their manufacturing plant is nearby, ask to take a tour. Some companies simply sell the dust that is left over from manufacturing other products. This may not have the necessary minerals for plants and trees, so it is best to buy rock dust from those companies that make it specifically for this use. They will be happy to provide a lab report you.

Rock dust is usually high in calcium, iron, magnesium, sulfur, and more than 100 other trace minerals. The PH level is often very high (eight or nine) and therefore must be used in small amounts, combined with compost, with peat moss if acid soil is required, or made into a liquid form. The rock dust does not dissolve when mixed with water, but forms a colloid, making it instantly available to the plants it is sprayed on. Using rock dust can replace many other natural mineral products that are harder to get and more expensive.

THE BACTERIA EAT FIRST

Not all the food that you add to your soil makes it to the plants. Initially, they must be broken down by soil microbes. The microbes live in many different areas of the soil, some live on the root hairs of many, some live only deep in the ground. The bacteria which live on the root hairs of plants convert minerals found in the soil, into a different form of the same mineral that is available to the plants. Microbes tear apart or combine minerals. Also when the microbes die they leave behind minerals in a changed form that is also available to the plants as food.

MIX THE DUST

For best results mix different sources of rock dust together to get complete trace minerals. There are many companies now offering rock dust. We have listed as many of them as possible in our Resource Directory. Make sure you add Kelzyme as it has the highest source of calcium!

MAKING YOUR OWN ROCK DUST

While this is not the easiest way, it can be done. Obtain a fifty gallon steel drum, weld bicycle gears at both ends. Hook up a bicycle through the chain gears. Set up the front wheel on a non-moveable base. Get a large round river stone to place inside. Make an opening that can be closed and locked. Then place local rocks into drum and exercise while crushing the rocks! Most soft rocks will work well. You can also crush lobster tails, clams, and other seafood materials. Add oyster shells to increase calcium.

HOW TO USE ROCK DUST

Compost Production: Using rock dust in compost production increases its energy level by adding minerals, and increasing the activity of bacteria. The increased bacteria stimulates the composting process. Rock dust also helps heat up the compost pile. Add a thin layer of rock dust to the compost pile, alternating between layers of grass clippings, manure, kitchen wastes, etc. Water as needed. As rock dust is high in PH level, use small amounts and add something acidic like leaves or pine needles. Never add chemicals to the compost. To lower PH level if it is too high, add one quart vinegar with five gallons water and sprinkle over pile once per week. See chapter on compost for more information. **Kelzyme** is good for this.

Pest Control: Rock dust is also very effective in controlling pests such as snails, because of its high silica content (67%). Dust lightly around the garden, allow to sit for 24 hours, then water well. Kelzyme is 32% Calcium! Very important in reducing stress, reducing pests!

On Trees: Trees also love rock dust. Always use small amounts as the PH level can irritate high acid based trees. A large tree should be given four coffee cans full of rock dust spread evenly beneath its canopy, starting two feet from the trunk to ten feet past the furthest reach of its branches. Sprinkle evenly then spray down with hose. Use a garden filter to filter out the chlorine from the water. See YBM in index.

In The Garden: Dust around the vegetable plants, allow to sit overnight, then spray down with filtered water. Mix into a liquid form and add seaweed. Spray this on the plants leaves, this increases the energy level of the plants, assisting them in fighting off pests. Use peat moss or composted aged wood to maintain a balanced PH level. Your vegetables will have increased mineral and vitamin content and will be more nutritious. Kelzyme is very good for the garden.

For the Lawn: When used on lawns , rock dust will increase color and encourage deeper root systems. Using rock dust will decrease your need to add any high nitrogen . Rock dust will also increase the effectiveness of compost applications. Add 50 pounds of rock dust per 1,000 square feet of lawn. Apply twice yearly. After dusting lawn, water well. Take care not to breathe in the rock dust when you are dusting your garden, lawn, trees, etc. Use Kelzyme on the lawn.

See page 47 and also Index for more info on Kelzyme.

SOME BENEFITS OF ROCK DUST

- Plants achieve physical completeness more quickly.

- Due to its high mineral content, rock dust helps in all aspects of the plants biological process.

- Plants are healthier.

- Plants are insect free. Less prone to attack.

- Plants can handle stress better.

- Plants live longer healthier lives.

- Rock dust is a natural fertilizer.

- Non polluting, biodegradable.

- Rock dust is inexpensive compared to other mineral products.

- Rock Dust increases mineral content of the soil and the plants growing on it.

SHOPPING FOR ROCK DUST

Some things to look for in a good quality rock dust. You'll want a very

Diatomaceous Earth

SOME SOURCES OF ROCK DUST

Contact your local state farming organization or local organic grower for local sources of rock dust or see index for IGA members.

SOME RESOURCES FOR ROCK DUST

See Resource Directory for Kelzyme, Nitron Industries, Arbico, Gardeners Supply, Peaceful Valley.

OTHER SOURCES OF TRACE MINERALS

ARBICO, Peaceful Valley Farm Supplies, Nitron Industries. Gardeners Supply. Kelzyme.

FURTHER READING

"**Remineralize the Earth**", 152 South St., Northampton, MA 01060; Newsletter, edited by Joanna Campe, 152 South Street, Northampton, MA 01060, USA (tel 413 586 4429), $18 subs. Research packet also available, $20.

"**Secrets of the Soil**", by Peter Tompkins and Christopher Bird;

"**Bread from Stones**", by Dr. Julius Hensel, available from Health Research, Box 70, Mokelumme Hill, Ca 95245;

Try **ACRES USA** as an excellent source of information and resources!

A video entitled "**Stopping the Coming Ice Age**" is available for $45 (UK VHS version) from People for a Future, 2140 Shattuck Avenue, Berkeley, CA 94704, USA (tel 415 524 2700). For the UK, specify the VHS version in your order.

Don Weaver and John Hamaker's book "**The Survival of Civilization**", $12 from Hamaker-Weaver Publishers, PO Box 1961, Burlingame, CA 94010, USA (tel 415 347 9693).

Harry Alderslade, 9 Walter's Row, Morrell Avenue, Oxford, OX4 1NT (tel 0865 240545), with an international network of remineralisation contacts.

Re: A new ice age from the greenhouse gases, contact **Ann de Vernal** of the University of Quebec and Gifford Miller of the University of Colorado.

fine dust, approx. 200 mesh. This is important because the finer the dust the more rapidly it is available to the plants and the microbes which have to eat it. The location it is mined in is important as it gives us a clue as to the forces that went into making it. High energy always makes high energy. Ask for a lab report. What PH is it? Don't use cement for construction use. Tell them what you want to do with it. Don't use any rock dust that has any type of chemical added to it.

The Test: A good test is to fill a clear glass half full with your sample and cover it with 3 inches of water. Shake it up and allow it to settle. The dust, silt and sand will settle into three different layers, with the dust settling on top. This will give you the percentage of how much of each you have. The finer the grind the easier the bacteria can get at it. However I have also found it to be true that small chunks, less then ¼ inch, are good for the soil as other creatures will eat it also. Another thing is that these chunks will break apart later providing additional food.

Another Test: Take any magnet and place on rock dust. If it clings to the magnet it is of the right energy polarity.

MAKING ROCK DUST MILK

Slowly add water to a cup of rock dust. Stir slowly until dissolved. Add to 1 gallon water, allow to sit, then strain into sprayer. This makes the food instantly available to plants sprayed. See Index for more uses of rock dust.

TYPES OF ROCK DUST

Rado Rock comes from Canada straight from the glaciers. Planters II comes from the Colorado Rockies. Earth Wealth comes from the San Gabriel Mountains in southern California. Azomite comes from Utah. Rock Phosphate is a well known rock dust, excellent when finely ground. New Jersey Greensand is also a very nice rock dust but can be expensive to buy depending on which side of the USA you live on.

A WORD OF CAUTION

Please realize that even though an item is organic and found in nature it can still be dangerous. Care should be taken whenever handling any formulas, chemicals or organic fertilizers, etc. Wash your hands, and follow the instructions carefully. Do not allow children to play with anything you make from this manual. It is better to be totally safe and sure than sorry later. Neither I, nor the company are responsible for damages incurred from using this manual incorrectly. I am providing you with Natural alternatives to chemicals, but nature's chemicals can be dangerous if misused. BE CAREFUL! Please let me know of your results and any questions. The information which is available to you in this book, is passed along with the understanding that we are always responsible for our actions, we must not upset the balance of our delicate ecosystem, nor endanger ourselves. We are always learning that nature is our greatest teacher and we as students must remind ourselves of this fact and be open to her lessons. As a race of beings we have much to learn. As the philosopher said:

"Don't cut the branch you are sitting on."

17

Your Own Vegetable Garden

THE VEGETABLE GARDEN is a very important link between us and the earth. As our garden grows we grow. When we grow our own food we know exactly what has been used to grow this food. We know that the fruit, the vegetables, the herbs that we grow are rich and full of nutrients necessary for our health. Our backyards can provide us with good homegrown food. Growing your own food is a very important aspect of maintaining a healthy mind and body. Even more important is growing our own food organically without the use of chemicals that pollute ourselves and our environment. If our home grown food is to have good nutritional value it must be grown in healthy live soil. The fact is, to succeed growing organically, healthy soil is a necessity.

Steps to Creating your Garden Paradise

RAISED BEDS OR NOT. THE DECISION TO HAVE A RAISED BED OR TO GROW IN THE GROUND DEPENDS UPON:

- condition of the soil

- amount of space you have to work with

- amount of time you have

- amount of water you have available to use

- amount of money you want to spend

A FEW POINTS ABOUT RAISED BEDS

1. Raised beds allow you to grow from 4 to 7 times more food in the same area then in the same ground.

2. Raised beds allow you greater control over watering costs, pests (from gophers to ants) and crops (especially extending harvest periods).

3. Raised beds allow you greater control of the soil being used.

Raised Beds are easy to make and you can recycle (reuse) various materials in making the beds. You can use rocks, wood (untreated, unpainted of course), tree logs, bricks. You can use bottles to make the sides of your raised beds. Stick upside down in ground. You can use clay on the inside to mold and hold them together or you can use sand and pack it in or use rope to tie them together. The idea has a lot of potential for reusing glass around the home. Untreated Redwood makes good wood for raised beds. A good size for a raised bed is 10 ft long by 5 ft wide by 12 inches tall with 2 in thick wood. This size is big enough to feed a family of 4. Pine, fir and almost any rather strong wood is all right as long as it is untreated. Never use Rail Road ties that have been treated. This is very bad for your health as well as being bad for the garden (it kills the soil).

If you decide not to use a raised bed then you must prepare the area first.

LAYING OUT YOUR GARDEN

Draw a plan for what you want to grow in your raised bed. Keep the tallest on one side with the small-

est to the opposite side. If you have more than one bed decide which plants will grow where. Change this design every year so as not to grow the same plants in the same place year after year.

Location: Choosing a location for your raised bed is very important. The location needs to be close to the kitchen to provide easy access for the cook. The location must provide at least 6 to 8 hrs direct sun, with the more the better. Must have proper drainage. So take a walk around your place and see if you can pick the perfect spot. Another consideration is water, it must have a close source.

Garden Filter: Never use city water in your raised bed. Many cities have chlorine (or ?) in their city water. Chlorine kills bacteria, that is what it does best. However an organic garden requires natural bacteria to function correctly. A garden filter or water transformer like YBM's (see appendix) will help keep your garden alive! You will notice an increase in worms and in the gardens overall health. An excellent Garden Water Transformer is available thru YBM. .

Putting together the Raised Bed: A good raised bed should be at least 4 ft x 10 ft x 12 inches high with 2 inch thick wood. Use non treated wood. The wood can be screwed together for easy break down when needed. If you have gophers in your area, you will want to screen the bottom with extra heavy chicken wire. The size of the bed depends on the area and amount of space you have to use. An ideal situation is to have two or more raised beds. One 4 ft x 12 ft x 12 inches high can produce enough food for a family of four. More beds allow you to rotate the beds and allow one bed to grow green cover that can be turned over.

Filling in the Raised Bed: The following should fit just right into the raised bed. You will have to use what you find in your area.

Start out with a good layer of old horse manure. To this I add either Llama pellets or Rabbit pellets (nature's time released fertilizer). Add 20 lbs of rock dust or any other trace mineral source. Add 20 lbs bone meal, 20 lbs alfalfa meal and 2 bales aged wood. Mix well. Water well (water slowly to allow soil to absorb). To this mixture add 500 lbs compost (if you have it) other wise add enough old horse manure to fill up to 4 inches from the top. Add another 2 bale aged wood or KRA wood product or any light soil. Add another 20 lbs rock dust, 20 lbs bone meal, 20 lbs alfalfa meal. The aged wood will insure the PH will be at the right place. A good PH for the garden is 6.5 to 6.8. Blend everything in together, watering as you go. Finally add enough mulch to fill the raised bed up to the top. Remember that this soil will settle after a few days, so keep a few bags of mulch handy to fill to top when needed.

COMPOSTING

The Secret to Growing Organically is Compost. Making good compost is a special art that we all must learn if we are to become Master Organic Gardeners.

Green Composting: 30% of our landfill materials come from this area of our wastes. Grass clippings, leaves, etc. make great additions to our composting system. A shredder will help to speed up the composting process.

Kitchen Composting: It pays to be able to save all of your kitchen wastes. A small container would be useful to have in the kitchen area for this purpose. Make sure it has a lid. We compost all of our kitchen wastes and recycled paper wastes as well as the wood from the matches we use, etc., any thing that will compost is saved. Learning to make compost is a great way to take something that is being thrown away and turns it into food for the soil, for the plants, and food for us.. This is recycling at its best! Make sure that you empty the container every day. I suggest that you can compost the kitchen wastes by layering into your compost. You can also bury the kitchen wastes into your garden allowing the worms a meal. Adding some rock dust to kitchen waste will help reduce smells and fly's, and will also increase microbial activity.

MULCH

Mulching is a very important part of the organic garden. It is an excellent way of recycling. It is always best to compost your mulch before using. Never mulch around plants with freshly cut mulch from trees or grass clippings. This will burn the plants.

What is the difference between mulch and compost?

Compost is the food and mulch protects the food from the elements like rain and the sun that will dry it. Mulch will hold water and not allow the soil/compost to dry. A good mulch is made from aged wood. This is wood that has been recycled from cut trees and composted organically. Often companies add urea to their compost believing that the compost needs this "chemical" nitrogen. So ask before you buy! Urea based products are very detrimental to the soils organisms and should not be used. Instead horse manure can be added. Rock dust will also work well here as rock dust will bind the nutrition together.

"Raised beds allow you to grow from 4 to 7 times more food in the same area then in the same amount of ground."

THE DRIP SYSTEM

A good drip system is important. A soaker hose will work very well here. The soaker hose can be buried about 2 to 4 inches from the top and can be moved as needed. Remember where the hose is to avoid damaging when planting. A battery timer will help to control the water and is easy to operate. Place a garden filter between the hose and the timer. Your garden will love this extra touch.

BUY EARTHWORMS FOR YOUR GARDEN

Earthworms will love your raised bed. Give them a head start by buying African red wigglers that are the best kind for this use.

Earthworms will love your raised bed.

PLANTING THE VEGETABLES

Start by making a list of the favorite vegetables that you like to eat. If you have not grown a garden before, I suggest that you take a gardening class to help you. Consider joining a garden club as they are a great source of not only help but seeds, resources etc.. The varieties that you choose to grow will be important and you must learn which will grow in your area. I also suggest that you get heirloom seeds instead of the commercial ones. Try Seeds of Change, Abundant Life, Native Seeds, Heirloom Garden Seeds, Native American Seed, Native Seeds/Search, Bountiful Gardens, Peace Seeds, Plant Finders of America, Ronniger's Organic Seed Potatoes, Seed Savers

Exchange, Southern Exposure Seed Exchange, For more resources try Acres USA. See Resource Directory for phone numbers and addressees to write to.

Mixing Flowers and Herbs

It is a good idea to plant flowers and herbs along with your vegetables for best insect protection. Try 50% flowers and herbs along with your vegetable garden.

Protection

A raised bed can be protected from extreme heat and cold by placing over it a sheet of plastic nailed down to the wooden sides. This will also help to allow new seedlings to grow and become established. Feeding. Use only natural organic fertilizers for best results. Chemicals will only destroy the balance of your garden. Allow time for this balance to occur. See making your own fertilizer chapter.

A final note:

One of my favorite things to do is to grow my own food. I really enjoy the whole process from starting a garden to deciding what to plant, what variety, where it goes in the garden and how to deal with all the various problems that may occur during a vegetable plant's life time. Someone told me the other day that you will have bugs and you will need chemicals to control them! I replied that while it is true that my garden will have bugs, it is not true that I will need chemicals to control them! What he was saying is that he did not know of any way to control them without using the chemicals. It has been my experience that if you believe in something it will happen and if you do not believe you will find a way not to make it so.

I have been fortunate in that unlike many people who started chemically then switched to Organics, I was never into chemicals and have always been organic in my approach to growing my food, my lawn, my flowers, etc. It is for this reason that my experience has shown me that it can be done 100% organically and that any problems are only waiting for the proper organic solution. It is even more important to grow your food organically since it is something that you must eat to be nourished by it. Chemicals have no place in our vegetable garden! Do not listen to the experts that tell you otherwise. Listen instead to your heart. What does it tell you? Do you poison your food as well as your mind?

Insects have a purpose in this world as do all things. Understand the purpose and you understand life! Who gave us the right to decide who lives or dies? Bugs are not pests but merely bugs living out their lives according to their genetic make up. They must do what they must do. Only we humans have a choice. Let us choose to live in peace with all things. Remember, Kill only as a last resort!

Provide for your garden all the tools it needs for bio diversity. Allow the many creatures of your garden to come together and sing their songs. Promote balance both within yourself and within your garden. Make the garden a place of life and not death. May your garden flourish and may you flourish!

"Allow the many creatures of your garden to come together and sing their songs. Promote balance both within yourself and within your garden. Make the garden a place of life and not death. May your garden flourish and may you flourish!"

Appendix

Appendix 1

When we study Organics, we also study the Science of Plant Toxicity. This toxicity works two ways; it protects the plants from the insects attacking them, and also protects them from being eaten by animals. As gardeners, we must be aware that many plants are dangerous when ingested. Small creatures such as cats, dogs and children are especially vulnerable. The young are in greater danger since they explore their environment using their mouths and hands.

It is not true that our pets will instinctively know which plants are OK to eat. If they were living out in the wild and had the benefit of a parent that showed them the way of the world from hunting to what plants not to eat, then they would probably not eat it, but with our modern day pets, many of Nature's Secrets lie hidden from them. Exposing our pets to these various exotic plants can be a learning experience that they will never forget and hopefully they will develop the correct habits for living together with us.

COMMON PLANTS THAT ARE TOXIC OR HAVE A BAD EFFECT ON YOUR PETS:

Oleander (Nerium oleander): Every part of this plant is toxic. This plant will kill a grown man if he eats only one leaf!

Datura: Highly fragrant and very poisonous. If the variety of Datura that you have has velvety leaves, this is a very poisonous variety that can cause damage just by rubbing against the leaves.

Castor Beans (Ricinus communis): One bite and you're dead. Very toxic. The leaves are toxic also.

Azaleas: Leaves if ingested in large amounts are deadly.

Rhododendron: leaves are toxic.

Boxwood (Buxus): all parts of plant toxic.

English Ivy (Hedera helix): fruit and leaves are toxic.

Heavenly Bamboo: contains cyanide.

Larkspur (Delphinium): all species of larkspur and delphiniums are highly toxic.

Ornamental Tobacco (Nicotiana): contains nicotine.

Oxalis: all species contains oxalates.

Pigweed (Amaranthus spp.): contains oxalates.

Spurges (Euphorbiacea): gopher purge, poinsettia are euphorbias and highly poisonous. Contains toxic amounts of oxalates that form crystals in the kidneys.

Poinsettia: can cause vomiting if ingested.

SOME VEGETABLES THAT ARE TOXIC TO ANIMALS.

Spinach, rhubarb, potato vines, onions and tomatoes may cause allergic reactions to your pets or may even be fatal to them. Bulbs such as **tulips, daffodils, amaryllis and iris** are dangerous if ingested. **Apricots and peach pits** contain cyanide and your pets should not be allowed to consume them.

ABOUT POISONOUS PLANTS

"Plant Poisoning in Small Companion Animals" by Murray E. Fowler, published

by Ralston Purina Co.

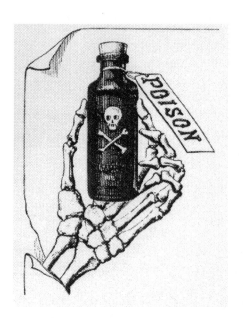

Appendix 2

THE GARDEN WATER TRANSFORMER

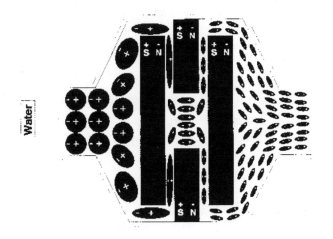

"At last... an easy way to grow plants the organic way, help protect the environment and conserve water at the same time."

The health hazards associated with chlorine are becoming better known to the general public. Most cities in the USA and around the world use chlorine in their water. Why? Chlorine kills bacteria and living organisms such as fungi, which contaminate our water sources and our oceans. Chlorine does a great job of killing these harmful organisms, but plants and soil rely on some of these fungi and bacteria to function.

Live soil is the cornerstone of the organic method. The soil must be alive with beneficial bacteria and fungi in order for the "system" to work. Food is eaten first by these bacteria, then made available to the plants through the soil. Every plant's survival is based upon receiving nutrition from the interaction between living organisms in the water and soil. Remove these 'friends' and we have big trouble! Many diseases originate in dead soil.

Many benefits of organic gardening are lost when chlorinated water is introduced into the ecosystem of the garden. Natural spraying depends upon using clean filtered water to obtain maximum results. An organic gardener or professional does not always have the time to solarize the water, so a good, long-lasting garden filter is important In this case its a water transformer using magnetic to change them chemicals at a molecular level to end up with chemical free "Rain" like water.

For more information on chlorine and its harmful effects, read "Nature & Health" magazine, Vol. 10, No. 4, Summer, 1989. Written by John F. Ashton and Dr. Ronald S. Laura. Pages 44 - 49.

The article is titled, "One Hundred Years of Water Chlorination", available from Whedon Young Productions, P.O. Box 170, Willoughby 2068, NSW Australia.

Until today, most gardeners had no choice of the water they used in their gardens. Now it is possible for all gardeners to steer away from chlorinated water and move toward a cleaner, healthier choice with YBM's Magnalawn 2000 YBM Magnetics Inc, 110 Terry Dr., Newton, PA 18940.

This water transformer is available directly from the manufactures at YBM (a commercial IGA member). 1-800-692-5296

Appendix 3

DIATOMACEOUS EARTH

Over 400,000 tons of pesticides are applied each year by American farmers with less than one-tenth of one percent actually reaching targeted pests! A main source of contamination of our soil, water, air and food, as well as being highly inefficient, this method of pest control places at risk the health of the farmer and consumer alike.

With the increase of organic methods of controls, we are seeing a reduction in chemical use. The farmer of today is beginning to understand the delicate balance of nature for which he is directly responsible. With this knowledge comes the search for safer, more effective methods of pest control.

Into the ring comes a long time favorite of organic farmers: DE. Throughout this book, I have been referring to Garden Grade DE (not pool grade). See DE in index for more uses.

DE (Diatomaceous Earth) contains 14 trace elements, is nontoxic and biodegradable. DE controls via a mechanical action rather than a chemical action. Most importantly, what this means is that subsequent generations of insects cannot become resistant to its effects.

DE STRUCTURE AND ITS FUNCTION

DE is derived from dead sea life called Diatoms. There are over 1,500 species of diatoms, and each has its own skeleton shape. Most are snowflake shaped, with a few being tube-shaped, and some cone-shaped. The shape is very important to its effectiveness. Diatom composition will vary according to the area in which they lived and died. Most DE contains silicon, sodium, aluminum, manganese, boron, magnesium, iron, calcium, and copper. The amount of amorphous (uncrystallized) silica the diatoms have present during fossilization determines the amount of crystalline silica found in today's DE. The greater the concentration of crystalline silica, the greater its effects of insects. Also, fresh water DE deposits contain lower percentages of silica than do salt water deposits.

HOW DOES DE WORK?

Look at DE under an electron microscope and you'll see that DE crystals (crushed Diatom shells) have very sharp edges with a large surface area (making DE very porous). Amorphous DE is very different from crystalline silica as far as health is concerned. Crystalline silica is hazardous to one's health, while garden grade DE (amorphous) is not. Crystalline silica causes silicosis (the World Health Organization has declared that crystalline silica levels in agricultural grade DE exceeding 3% can be dangerous to humans). This is why pool grade DE is not safe to use in your garden or farm. Pool grade DE is made by superheating amorphous DE until it becomes 100% crystalline silica! Many companies add an acid to complicate matters. Avoid using pool grade DE at all costs!

DE causes insects to lose their body water content by more than 10%. DE affects the insect's natural coating of cuticle (a waxy coating secreted by the epidermis). The sharp edges of the DE particles lacerate this coating, penetrate between the insect's body plates and absorb any moisture with which it comes into contact. This slowly dehydrates the insect. The DE affects the insect from the outside as well as from the inside if the insect also eats the DE.

DE can be fed to your dogs, cats and horses without any harmful effects. It is a natural dewormer, and also provides trace minerals. Nitron Industries, Arbico, Gardeners Supply, are good sources of an excellent quality DE.

A good formula to follow for feeding is:

One tablespoon per daily feeding for your larger dogs (over 50 pounds). One teaspoon per daily feeding for your smaller dogs. 1/2 teaspoon per daily feeding for your cat. One cup per daily ration for your horses. DE can also be added to their drinking water at 1 tablespoon per gallon water or 1 cup per 50 gallons water.

By feeding DE to your animals, you are also providing for natural fly control since DE will reduce fly populations. This happens because DE travels through the animal's digestive system, and is deposited, unaffected, in the animal's feces. The flies deposit their eggs as usual in the manure but few larvae will survive because they must move through the manure in order to feed. DE can also be eaten by people. I suggest that you take a teaspoon in your meals once a week.

Garden grade DE is classified as GRAS(Generally Recognized As Safe) by the federal government and has been exempted from the requirement of residue tolerance on stored grain. So far, health professionals have considered Amphorus DE not to be a cause of cancer in animals and humans.

A Word of Caution

While DE is safe to use, it would be wise to follow a few safety rules:

· Avoid inhaling; use a face mask if you have a breathing problem.

· Avoid contact with the eyes. Wash with water asap. Do not rub !

· Avoid overuse. DE will kill beneficials as well as earthworms if too much DE is applied to the ground.

While DE does a great job of protecting plants from insects, it will do a better job if the overall coverage is increased. This is best done by using an Electrostatic Nozzle developed by PermaGuard of Albuquerque NM. This device charges particles by forcing them through an ionized field, where they pick up negative electrons. Surfaces that are attached to the ground are positive and will attract the DE dust. And due to the fact that like charges repel, when one area of a surface is covered, it repels oncoming particles, forcing them to move over and find an unused place. Eventually, the dust will cover the entire leaf surface, both on top and underneath. This makes for a very effective use of DE. See index for more uses for garden grade DE.

Some Sources of DE:

DE DISTRIBUTORS AND MANUFACTURERS

<u>Nitron Industries, Inc.</u>

4605 Johnson Rd.

P.O.Box 1447

Fayetteville, AR

72702-0400

1-800-835-0123

<u>Peaceful Valley Farm Supply</u>

P.O.Box 2209

Grass Valley, CA 95959

1-916-272-GROW

<u>ARBICO</u>

P.O.Box 4247

Tucson, AZ 85738

1-800-827-2847

ELECTROSTATIC **DE** APPLICATOR

<u>Perma-Guard, Inc.</u>

3434 Vassar NE

Albuquerque, NM 87107

Use The Duster Miser to Apply DE in your garden. Try Nitron or Arbico or Peacefull Valley.

Appendix 4

SAFE SOAPS

Dr. Bronners Soaps

P.O. Box 28

Escondido, CA 92025

All-One-God-Faith, Inc

1-619-743-2211 and in all health food stores

Amway's LOC

7575 Fulton St., E

Ada, MI 49355-0001

Amway Corp.

Available only through local distributor:

1-616-676-7948

Citrus Organic Cleaner

Natural Bodycare

Available in most stores

Citra Solv

Citra Solv

Available in most stores

Jungle Rain

Chris Klein

953 Sidonia St

Encinitas, Ca 92024

1-619-436-6605

Herbal Soaps

P.O. Box 106

Altadena, CA 91001

Erlander's Natural Products

213-797-7004

Coconut Oil Soap

3920 24th St.

San Francisco, CA 94114

Common Scents

415-826-1019

NOTICE

Address and phone is likely to change. We are not responsible for any such changes. If you are an IGA member, contact IGA for new address/phone.

This is a list of only a few sources of natural soap products. Read the ingredients before buying. Experiment with the various organic soaps to use for pest control. Remember that you are defeating the purpose of growing organically if you use a soap with polluting chemicals in it.

Appendix 5

PREPARATIONS OF FORMULAS

Garlic Juice from Cloves: Take cloves and crush (either run through juicer or use garlic press). Use 1 bulb (which has many cloves attached) per gallon of distilled water (or preferably solarized). Crush each clove and allow to sit in gallon water over night. Run through strainer. You can also make an infusion from the leaves.

Solar Tea: Solar Tea is made by placing the ground up dried leaves of the herb into panty hose tied into a ball (becomes a tea bag) then place into a gallon (or larger) glass container of pure water (not city water, distilled or filtered water is best but stream water is OK). You should always use a water filter to control unwanted bacteria, toxins, and/or chlorine). Allow to soak in the direct sun and moonlight for 24 hrs. Pour into a clean dark glass container. Use within the next 24 hrs. The strength of this mixture will depend upon the purpose and the herb used.

What part of the plant are you using?

Leaves: Use dry leaves, grind in mortar and place in panty hose, tie into a ball and allow to soak for 24 hrs, use large gallon (or 5 gallon) glass containers only, use only once.

Bark: Allow bark to dry on sun tray, then grind bark to powder, place in panty hose, tie into ball and add to container of water and allow to soak for 24 hrs, strain through filter.

Seeds: Grind seeds with coffee grinder, place into panty hose, tie into ball and allow to soak for 24 hrs. Filter before use.

Flowers: Pick flowers, allow to dry on solar tray then grind into powder with mortar, place into panty hose and allow to soak for 24 hrs.

Fruit: The fruit can be either dried or used fresh. Dried... place fruit on solar dryer, allow to dry slowly and grind to a fine powder. Place into panty hose to soak. When using fresh fruit, it is best to use a blender and liquefy the fruit for easier application.

Root: Can be used two ways, fresh or dried. When used fresh, can be boiled and a thick mixture can be made by allowing to soak for 24 hrs over slow heat. When dried, can be added to panty hose to soak for 24 hrs.

SOLARIZATION OF WATER

Whenever you are preparing a spraying formula, always use solarized or filtered water. Solarized water has been allowed to sit in the sun for several days. The mixture should be stirred clockwise for 5 minutes then stirred for 5 minutes the other way. This should be done once every day the liquid is being solarized. Use a glass container if possible. Colored glass will change the energy level of the liquid. Experiment for best results. Green glass will be very helpful for promoting new growth on plants while yellow for new buds and blue for fruit. Red is useful for repelling insects.

HOW TO PREPARE HERBAL TEAS FOR SPRAYING

Method 1: Empty herbs into a large pot of boiling water. Cover and steep for 1/2 hr. Pyrex or Corningware are preferred containers. No aluminum or cast iron. One tablespoon per cup of boiling water. Strain before using. Can be used straight as a spray etc. Excellent formula to use for repelling bugs and diseases.

Method 2: Slowly bring water to a boil (use low flame), then turn off heat, add 1 cup of the herb, steep for 5 minutes, turn on heat to low and simmer for 5 min. Turn off heat and allow to sit overnight. Use strainer to filter and pour into clean glass container. Label. Use within 24 hrs. Makes a very strong concentration. Excellent for making a fungus control spray or ideal for that hard to kill pest.

MAKING A SLURRY.

When you are using a dust or herb that is very dry, it will not blend well with water at first. Therefore a slurry is made. This is done by half-filling a cup with the material you want in slurry form. Add 1 drop of Shaklee's Basic H or any other natural soap. Slowly add water while stirring. The mixture will accept water and dissolve into a paste-like substance. This can be then added to water and used as a spray. I suggest adding to 1 quart water, and dissolving by shaking well. Save this quart and use 1 cup per gallon liquid to be sprayed.

Appendix 6

DR. BRONNER'S ALL-ONE-GOD-FAITH,
A 'SOAP AS MESSENGER' SUCCESS STORY

By Heather Allison-Jenkins

Readers please note: This article has the distinction of having been deemed the BEST article ever written to date about the good Dr. Bronner (so say Jim and Ralph!) ...and believe me there have been hundreds of pieces put into print. Printed with Ralphs blessings......Andy

Joining the ranks of natural soapcrafters takes heart, dedication, imagination, and perhaps a desire to leave this world at least a little better than when we arrived - if at all possible. Imagine a man, an immigrant to America, who has been believing in his pure castile soaps for nearly 50 years, and who like his fellow soapcrafters takes great pride in bringing to the public his simple and pure products. Imagine the success of this one man - a humble soapcrafter like you and me - growing to millions of dollars in soap sales each year. Now imagine if you can that this man's soap is not his life's work. It's the label.

If you have ever read the label on a bottle of Dr. Bronner's Pure-Castile Soap, you surely remember doing so - even if it happened decades ago. You may have said to yourself, "Whoever wrote this had to have escaped from a loony bin," which is entirely true, but at a closer glance the apparent madness of Emanuel Bronner, 88, of Escondido, California, is in fact complete sanity. While most of us soapcrafters who sell

our precious soaps are doing what we can to help our Earth by donating things to peace-loving causes, Dr. Bronner has concocted his own peace plan. The 3,000 plus minuscule words typeset in all directions on his labels tell about it, his "All-One-God-Faith", in which he hopes to unite all mankind on "Spaceship Earth".

"The whole world is our country, our Fatherland, because all mankind is born it's citizens! We are all brothers and sisters because One, Ever-loving Eternal Father is our only God and All-One-God-Faith reunites God's legion! LISTEN CHILDREN ETERNAL FATHER ETERNALLY ONE!" Thomas Paine

"We're ALL-ONE or NONE! ALL ONE!" The style of this is consistently sparse and direct. Much of Dr. Bronner's philosophical statement is inspired by the ancient scrolls of every religion, and the lives and words of many great people living from before Christ to the present. What's in the bottle is just as impressive. "Pure Castile Soap", "No Detergents", "No Synthetics! None!", "100% Cruelty Free!". The ingredient list is short and sweet, and only the finest pure oils are used for scent and sensation.

Dr. Bronner (a self eschewed Master Chemist and Essene Rabbi through his life's own experiences) has according to his sons Ralph, 60, of Menomonee Falls , Wisconsin and Jim, 58, of Escondido, California, who run the business now with 20 wonderful and loyal employees, never stopped revising and honing his label and the message it carries. "He has spent all of his life working on it (the label)," says Ralph, of his father.

"My father in his day was so intense: 'Ralph, you're not my son unless you memorize this label.' Which one? Each one was 3,000 words, and each one was different." (There is a lot of repetition, but each label is different on each type of soap. Peppermint, Almond, Baby Super mild Unscented, Eucalyptus, Lavender, and Sal Suds Organic Cleaner.)

Jim and Ralph are proud to be fourth generation soap-maker masters. Their father learned the trade from his grandfather in the 1920's in Germany. Dr. Bronner arrived in the States in 1929, and bounced between soap consulting jobs for various companies. Imagine the shock to the family when he announced his intention to marry a catholic hotel maid! This love dynamic in his life was the catalyst for his search for religious truth. By the late 40's he was mixing up tubs of his soap with a broom handle in the Los Angeles hotel room he was living in, and selling it after his lectures about his peace plan and the "FULL TRUTHS" he had come to understand could unite the entire world. It was about that time that the idea presented itself to place what he calls his "Moral ABC's" - a complete version of this is now available in book form - on a label.

Unfortunately, Dr. Bronner's Moral ABC, which dealt with everything from Spaceship Earth under One God, anticommunism, ecological ideas, etc. was far ahead of its time and was not understood by many. Through a series of tragic events he was forced into an Elgin, Illinois mental institution against his will in 1946. Everything from electric shock to rigid confinement was tried in order to "turn him into a normal person." Luckily, after six months and two failed escape attempts, he succeeded in escaping to California without a penny and where no one knew him. After sleeping penniless on the roof of a YMCA with the pigeons, he finally earned some money fighting forest fires as a volunteer. Here he started to make his soaps and food products and use them to help spread the Moral ABC. The soap became a messenger for his ideas.

Since then, he has acquired a sort of cult following which surged in the 60's with the peace and Earth-loving hippies and has continued on into the 80's and 90's with New Ager's buying the bulk of the over million bottles hand packed annually. It is also spreading among mainstream America and can be found in outfitters supply stores and in every health food store. In New York City it is highly prized by models and is found in the finest cosmetic stores. Dr. Bronner himself has become somewhat of a guru, and has been bestowed with the unofficial title of, 'The Pope of Soap'. Many fans have made the pilgrimage to his California home - he's not hard to find, his address and phone number printed on every label . Though the good Dr. suffers today from total blindness (since 1974) and advanced Parkinson's Disease, the main office for the company is still in a room in his home.

The Bronner's have used part of their profit for worthy causes including: buying buses for a local Boys' and Girls' Club, a well in Ghana, Africa to provide fresh water to the natives, reforestation projects, offering scholarships for students, and so much more. The All-One company also has a wonderful profit sharing program for the workers including complete medical, optical and dental coverage. Other people may not understand Dr. Bronner's obsessive need to spread the wealth he's been blessed with, but it is because he practices what he preaches. His life and his incredible energy, outlook and drive can bee seen in the statement he made to his son Ralph years ago, "I'm the luckiest blind man in the world."

"What is beautiful about my father is that he believes everything on that label," says Ralph, who says that way back when, he suggested his father's soap - of which the uses range from shampoo to toothpaste, laundry cleaner to bug repellent, and plant cleaner to body message (though customers swear there are hundreds more uses) - something catchy like 'Mint Glow'. "But my father wouldn't hear of it. That turned out to be a good thing." Since then, the famous label has been called by admirers, devotees and some of the hundreds of mainstream publications who have asked for permission to write stories about Dr. Bronner's soap, "The most incredible piece of literature I've ever seen printed on a six-by-six piece of paper," and "The power of word over design."

One Swedish newspaper called Dr. Bronner's. "The soap that crossed America on quality alone," and it's true, the Bronners have never spent a penny on advertising or marketing their soaps, but have relied solely on word of mouth and unsolicited articles for their word to be spread. It will probably never be clear whether it is the soap or the message on the bottle which has made Dr. Bronner's more than a product, but a symbol of an ideal, and an ideal product. The Dr. himself might say it was faith in his loving God of all mankind and all the good Karma accrued from fighting the good fight. Whatever the reason, Dr. Bronner's soap business - though entirely unique and induplicable - is testament to the fact that upstanding ideas and pure and natural ingredients can indeed succeed. Dr. Bronner will leave this world a better place. He is an inspiration to soapcrafters everywhere.

Dr. Bronner's All-One-God-Faith

P.O. Box 28

Escondido, CA 92033

1-619-743-2211

Ralph Bronner

W172 N9335 Shady Lane

Menomonee Falls, WI 53051

1-414-255-5680

Appendix 7

Primordial Soup

Organic Gardening depends on having a living soil. The process of assimilation, adsorption and growth is a living system is dependent on bacteria. Bacteria forms the key corner stone for the organic process to happen. I learned early on in my childhood training (my mother and grand mother are both organic gardeners and taught me all about bacteria!) that bacteria is a very useful tool in Organic Gardening; from compost tea to producing the various biodynamic sprays.

SuperSeaweed™ which is an invention of mine, is a result of many years of experimentation and assimilation of my mothers knowledge as well as from my own experiences as a gardener. I like to call SuperSeaweed a "microbiological activator". In simple terms, it is a special blend of five different types of deep ocean seaweed, chosen for its purity and for its special bacteria. Each seaweed, being from a different part of the world, contains its own special bacteria and its own special minerals. Blending these different liquid seaweed's produces a complete bacterial and mineral 'soup'. To this mixture I have added rock dust, which increases mineral content such as calcium, iron, etc. I also add activated Willard Water™, Nitron A-35™, and Agri-Gro™, Acadie™.

When using **SuperSeaweed**™, only 5 capfuls per gal is required for regular feeding. SuperSeaweed is not a fertilizer but acts as a 'vitamin pill'. SuperSeaweed also helps to promote needed bacteria and enzymes when used with every watering. Superseaweed allows the soil and plants biological systems to begin to work. Plants and soil require composting and applications of mulch to insure a complete and balanced system.

SuperSeaweed™ is an excellent Rx for sick plants. Use 10 capfuls per gallon water to start with. This will give your plants a boost and help them along in the recovery. Allow a week for results. Then reduce to 5 capfuls per gallon. Repot soil if needed.

SuperSeaweed is excellent for fruit trees and vegetables. Use 10 capfuls per gallon, add to sprayer and spray on leaves for fast results! **Superseaweed** is an excellent foliar feeder for your roses! Use 10 capfuls per gallon, add to sprayer and spray regularly on leaves. See rose chapter for more info on how to use SuperSeaweed™.

Use **SuperSeaweed**™ on sick trees! Use 10 capfuls per gallon, add to sprayer and spray leaves on trees. Spray weekly for a month. Trees take longer to respond. Soak with **SuperSeaweed** around base of trees. Pour into tree vents. See Tree vents in index. SuperSeaweed™ works with in-line feeding systems as well as with hydroponics systems. To increase effectiveness add equal amounts of molasses (i.e.. 1 capful **SuperSeaweed**™ = 1 capful molasses).

SuperSeaweed™ works great on orchids! Most any flowering plant will do well with regular additions of SuperSeaweed to its water. Since **SuperSeaweed** is not a fertilizer, I suggest that you add it to your regular liquid organic fertilizer. **SuperSeaweed** makes your fertilizer last longer and you will have to use less. **SuperSeaweed**™ also acts as a wetting agent. This allows liquids used in fungus control to work better and should be added to all fungus sprays. **SuperSeaweed** was invented by myself in 1972 while I attended school at the University of Florida located in Gainesville Florida. It was made to be used in conjunction with other organic fertilizers and not with chemical fertilizers. If you must use a chemical fertilizer, use 1/2 as much of the chemical fertilizer while increasing your organic fertilizers. To Order please see Member application on order page.

Appendix 8

ANDY LOPEZ - THE INVISIBLE GARDENER

Living in the secluded hills of Malibu Canyon is a man known as the "Invisible Gardener." With a title like that. some confusion is inevitable: Could he be a playful, elf-like man who comes down from the canyon at dawn to sprinkle dew over the ground before others awaken? Or simply a wonderfully unobtrusive groundskeeper who literally fades into the landscape while dutifully taking care of the foliage? Actually. as Andy Lopez explains, *he* is not the 'Invisible Gardener" at all - nature is. 'I am just one of her helpers,' he states. "And if I could get more people to do what I am doing. instead of consistently destroying the environment she would have a much easier job." Growing up in Puerto Rico/Miami, Lopez was heavily influenced by the fact that his mother grew her own fruits and vegetables and always used animal manure as fertilizer This organically based philosophy toward gardening techniques laid the foundation of Lopez beliefs and in 1972 he founded Astra's Garden, based on something of a religion that subscribes to living in harmony with the environment: not polluting; treating all living plant life with respect; and, basically just listening to what the earth is telling us.

People call him a "soil psychologist" yet he prefers the "plant nutrition specialist" appellation instead. Still he does make an effort to get to know his clients - to get a sense of their Lives - as much as possible. "I try to deal with the owner and his/her property as one

entity he maintains, going onto state that it is remarkable how much one's property reflects one's emotional state. His wide variety of clients - including in the past, such celebrities as Olivia Newton-John and Mark Harmon, attests to his effectiveness in connecting with people who share his motto ("happy, healthy, holy") when it comes to living and interacting with the natural surroundings.

Lopez is also the founder of the Invisible Gardeners of America (IGA) club. which he started in an effort to raise the consciousness of all those concerned with their environment The club publishes a newsletter every month on website or yearly via snail mail, which discusses new products new procedures, and a number of timely topics relating to planting and growing. It also produces a constantly updated compendium of current environmental reports. providing an essential tool toward understanding the battle against - and alternatives to - polluting chemicals, which is available for free to IGA members on his website.

Lopez professes to be the only 'Absolutely Organic' spraying/pest specialist in the United States (at this time, even though he expects more to come out), pointing out that. while others might be semi-organic, they all still rely on chemicals in one form or another. "If It doesn't come from mother earth, then it isn't organic," he states flatly. He goes on to say that 75 percent of all pollution in the U.S. comes from the home: If people would go back to the basics and grow their own food, and take care of their bug problems naturally, chemical pollution would no longer be a problem. Incorporating the philosophies of his various organizations and activities, Lopez's mission is the education of the public on the necessity to quit using chemicals altogether and in the meantime, to properly dispose of their waste - so that one day the earth may return to its natural cycle of growing and decomposing. He also professes hopes of expanding his club internationally noting that. "After all, these same problems exist all over the world. They just exist in varying degrees".

R e s o u r c e   D i r e c t o r y

The following pages are IGA Commercial members. They have joined IGA and passed our rigid testing. We highly recommend these companies to you. Know that you can trust them and their products. Tell them if you are an IGA member (have IGA member number ready) and see if they will give you a 10% discount, but surprises are nice! We attempt to keep their phone numbers and other info up to date but these things change so fast so that it is possible to get a wrong number, if this happens please contact our office in Malibu, CA (m-f 9 am to 5 pm) and we will provide you with the correct information. If you use the internet try our website first since we always post new info here first. We also accept faxes and e-mail.

Want to get listed? We list only our IGA members.

IGA commercial membership fee is $100 per year. What do you get for this? You will be in the next printing and in every printing after that for as long as you renew your yearly membership fee. You will get a new book sent to you. Each time we print we print more books and get a wider distribution of our books. Currently available world-

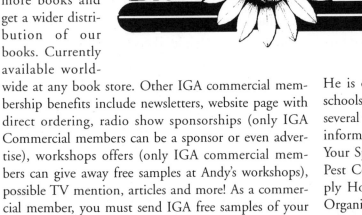

wide at any book store. Other IGA commercial membership benefits include newsletters, website page with direct ordering, radio show sponsorships (only IGA Commercial members can be a sponsor or even advertise), workshops offers (only IGA commercial members can give away free samples at Andy's workshops), possible TV mention, articles and more! As a commercial member, you must send IGA free samples of your products for us to test and promote. Andy will be happy to write an article, recommendation of your products. Drop us a line and we will send you our membership application or see order form in this book.

The Invisible Gardeners of America

P.O.Box 4311 Dept. NPC98

Email: cmemb@invisiblegardener.com

Please make a copy of this page when ordering.

1-310-457-4438

http://www.InvisibleGardener.com

About Andy Lopez

THE INVISIBLE GARDENER

Andy Lopez, The Invisible Gardener, has been in the organic business since 1972. While attending the University of Florida in Gainesville, he started The L.I.S.T (The Living Institute of Survival Technology), The Gainesville People's Yellow Pages and The Florida Pages, a new age travelers Guide to Florida. While in Florida he started the Miami Organic Growers Club, attended on the board of the Rare Fruit Council, the Rose Society, the Horticulture Society to name a few. In 1972, he developed Organa™ (the SuperCompost), Organa Gold™ and Malibu Gold™ as well as a seaweed concentrate called SuperSeaweed™. In that same year he started the First 100% Organic Spraying and Pest Control Service in the USA. Using natural every day materials he developed a unique system of fertilization and pest control designed to increase balance.

Moving to Malibu, California in 1984, he brought with him his company and has been here since.

He is currently teaching at many organizations and schools. He has several radio shows and has produced several videos. Contact The Invisible Gardener for more information. Author of " How to Heal the Earth In Your Spare Time" and recently his new book "Natural Pest Control". The Invisible Gardeners Organic Supply House is your source for hard to find Advanced Organics. We make our own compost Organa™, Organa Gold™ and Malibu Gold™ and of course SuperSeaweed™. Call us for your needs, Organically! Services provided by The Invisible Gardener: Natural Pest Control Services, Natural Tree Care Services, Organic Consultant Services, Organic Spraying Services, Classes, Workshops, Videos, Tapes! We do house calls! Visit our website on the internet!

HAPPY GROWING, ORGANICALLY OF COURSE!

Andrew Lopez
The Invisible Gardener

Agri-Gro has been my weapon of choice in battling diseases and pests. Now they have even more products available for you to use on your lawn and garden. This stuff is amazing on trees, fruit trees and roses!. I highly recommend this product also if your property is new and you want to clean it out of previous chemical use. Tell em The Invisible Gardener sent ya!

Ronald Smith
HC 4, Box 333
Doniphan, MO 63935
1-800-881-8801

If you want to be independent of the electrical/ social grid then you must use this company! The state of the art tools in solar living, wholelistic lifestyles. Ask for their Solar Living Sourcebook..from photovoltaic, wind and hydro powdered systems to super efficient refrigerators, composting toilets, etc.

1-800-762-7325

This company is a pioneer company in Orgaics! Their Nitron A-35 is the finest enzyme product on the market! They also are a great source for organic fertilizers, rock dust, compost, and they also carry the best DE around! Ask about their Natural Meal, Fish Emulsion (without UREA!) and much more. Tell em Andy sent ya !

Nitron Industries
1-800-835-0123

This company makes one of the nicest organic fertilizers I have ever used! My trees really like this stuff! I highly recommend using with tree vents, roses, fruit trees. What am I talking about? **BioSol** of course! They are also an excellent source for various ground cover seeds from turfgrass to legumes to native trees(ca) and wildflowers! See more info on Biosol in index.

1-800-423-8112 Southern Calif.
1-800-339-8245 Northern Calif.

Deer Off is a safe effective way to control deers and other creatures from eating up your roses and other ornamentals. Works for months! Will keep them off lawns too! I also use it for gopher and squirrel control.

Tell em The Invisible Gardener sent ya!

**Deer Off
1-800-DEER-OFF
http://www.deer-off.com**

This is where to send your soil sample s to be testing. Ask for your Organic Report. Also ask about their Magic Garden Kids Program. See page 166 for more info on this service. Tell 'em The Invisible Gardener sent ya!

**1-800-431-SOIL
http://www.greengems.com**

This is another Organic Pioneering Company! One of the Worlds Oldest Commercial Insectary! Lots of good info here plus an excellent source of beneficials! From Decollate Snails to Fly Parasites and much more! Tell'em Andy sent ya!

**Rincon-Vitova Insectaries, Inc
P.O. B ox 1555,
Ventura, CA 93002
1-800-643-5407**

email bugnet@west.net

**P.O. Box 8800
Metairie, LA 70011-8800**

I f you want to be on the cutting edge of eco-farming technologies, techniques, markets, news, analysis and trends, look to Acres USA. These folks will tell you the truth as to whats going on in Organics!. Never miss an Issue, I know I don't! Subscribe today! mention IGA for your free sample issue of Acres USA! along with latest book catalog!

1-800-355-5313

All Natural
PEST CONTROL

Andy Lopez (The Invisible Gardener) is world recognized as an environmentally concerned horticulturist and has been in the field of Organics for over 29 years. Founder and CEO of The Invisible Gardeners of America. Andy has developed unique products and systems to naturally sustain your home and garden in perfect harmony with the environment.

YOUR LIFETIME MEMBERSHIP BENEFITS INCLUDE:

☑ New members receive free book " **Natural Pest Control- Alternatives to Chemicals for the Home and Garden"**, the latest book written by Andy Lopez The Invisible Gardener.

☑ One Newsletter per year via snail mail or visit IGA website for free monthly **Online Newsletters** plus all IGA members get F**ree Digital Newsletter** emailed to them once per month.

☑ Free unlimited consultation with Andy via **The Organic Hot Line**(tm) m-f 7-9 pm Ca time

☑ Free unlimited use of **The Organic InfoBank**(tm) website. IGA members only. ID and Password required.

☑ New IGA members receive **1 Free Online Class** via snail mail, fax or internet. General Public $20, IGA members $5 per class.

☑ Special prices on IGA sponsored workshops and classess.

☑ Special prices on compost and organic supplies.

☑ IGA membership required for all local services: natural pest control, natural tree care, natural spraying services.

☑ Best of all you will have a 100% organic property!

The Invisible Gardeners of America
P.O.Box 4311
Malibu, CA 90265
1-310-457-4438 office
1-310-457-5003 fax
http://www.invisiblegardener.com

IGA ORDER FORM

Date	
Name	
Address	
City, State, Zip	
Phone Number	
Fax Number	
Email	

QTY	DESCRIPTION	UNIT PRICE	TOTAL
	Individual Lifetime IGA Membership	**$55.00**	$
	Senior Lifetime IGA Membership	**$35.00**	$
	Non-Profit	**FREE**	**FREE**
	IGA Commercial Member	**$100 per**	$
	IGA Subscription Service	**$25 per**	$
	Quart Superseaweed	**$45**	$
	8 oz SuperSeaweed	**$12**	$
	90 min Las Vegas Workshop Video	**$29.95**	$
	90 min The Healthy Garden Show Video	**$29.95**	$
	60 min Dances with Ants Video	**$24.95**	$
	Recent 60 min radio tape of Andy	**$12**	$
		SUBTOTAL	$
	No taxes on IGA membership or subscription service..........CA SALES TAX (7.50%)		$
	Add all shipping charges and put here...........................SHIPPING & HANDLING		$
		OTHER	$
		TOTAL	$

Index

Tachinid Flies 50
Tangelfoot 47, 104, 113
Tansy 60
Teas 60
Tent Forming Caterpillars 115
Termite Control 92
Termite Prevention 89
TERMITES 89, 92
TERRO 13
thrips 59, 63
Thyme 60
Tobacco 32, 47, 52, 60, 112
Tomato Plant 61
Top Dressing 96
trace mineral spray 121
trace minerals 130
Traps 48, 65
Tree Vent Formula TGM+ 112
Tree Care 111
Tree Decay 115
Tree Tea Oil 57
Tree Vents 111, 113
Trees 111

V

Vacuum 77
Vacuuming 84
VEGETABLE GARDEN 133
Vegetable Oil 73
Vegetable Organics 51
Vegetables 135
vine crops 71
Vinegar 8, 116, 121
Viruses 70, 115

W

Wasp 50
Water Control 90
Water Soaked Soil 70
Water Transformer 65
Watering 71, 98, 100
Watering Methods 46
Watering Procedures 95
Weed Control 98
Weeds 70
Wetting Agent 122
White Fly 65
whiteflies 59, 63
whitefly predators 65
Wild Flowers 61
Willard Water 120
Winter Injuries 116
wood soil conditioner 94
Wormwood 61

Z

Zinc 72

Order Form

If your favorite retailer or catalog is sold out of *Natural Pest Control,* use the following order form.

Telephone orders: (800) 265-4367 or (419) 281-1802

Fax orders: (419) 281-6883

E-mail orders: orders@bookmasters.com

Mail orders: Use this form for orders paid by check, Visa, MasterCard or American Express. Make checks payable to: Bookmasters, 1444 U.S. Route 42, Ashland, OH 44805, USA

Inquiries: Telephone (707) 823-1999

Price

$19.95 per book. Please add 7.25% sales tax, or $1.45 per book, for orders shipped to California addresses. Ohio orders add 6% sales tax or $1.20 per book.

Shipping and handling

$3.50 per book US, $5.50 all other countries

Also available from Harmonious Technologies: *Backyard Composting,* $6.95 plus shipping and handling fees of $3 US, $4 all other countries.

Payment

_____ Check _____ Visa _____ MasterCard

Card number _____

Name on card _____ Expiration date _____

Number of books ordered _____ Amount enclosed _____

Company name _____

Name and title _____

Address _____

City _____ State _____ Zip_____ Country _____

Telephone _____

E-mail address _____